Transboundary Water Man
and the Climate Change D

Climate change has an impact on the ability of transboundary water management (TWM) institutions to deliver on their respective mandates. The starting point for this book is that actors within TWM institutions develop responses to the climate change debate, as distinct from the physical phenomenon of climate change. Actors respond to this debate broadly in three distinct ways – adapt, resist (as in avoiding the issue), and subvert (as in using the debate to fulfil their own agenda).

The book charts approaches which have been taken over the past two decades to promote more effective water management institutions, covering issues of conflict, cooperation, power, and law. A new framework for a better understanding of the interaction between TWM institutional resilience and global change is developed through analysis of the way these institutions respond to the climate change debate. This framework is applied to six river case studies from Africa, Asia, and the Middle East (Ganges-Brahmaputra, Jordan, Mekong, Niger, Nile, Orange-Senqu) from which learning conclusions and policy recommendations are developed.

Anton Earle is Director of the Stockholm International Water Institute (SIWI) Africa Regional Centre based in Pretoria, South Africa.

Ana Elisa Cascão is a Programme Manager at SIWI's Capacity Building services and Co-director of SIWI's international training programmes on Transboundary Water Management for Lake Victoria, the Middle East, and the Tigris-Euphrates.

Stina Hansson is performing research on transboundary water management and water services provision at the School of Global Studies, Gothenburg University, Sweden and at SIWI.

Anders Jägerskog is Counsellor for Regional Water issues in the MENA region at the Embassy of Sweden in Amman, Jordan. He was formerly Head of the Transboundary Water Management Unit at SIWI where he also served as the Programme Manager of the UNDP Shared Waters Partnership (SWP).

Ashok Swain is a Professor at the Department of Peace and Conflict Research and at the Department of Earth Sciences of Uppsala University, Sweden.

Joakim Öjendal is a Professor of Peace and Development Research at the School of Global Studies, Gothenburg University, Sweden.

Earthscan Studies in Water Resource Management

Hydropower Development in the Mekong Region
Political, Socio-economic and Environmental Perspectives
Edited by Nathanial Matthews and Kim Geheb

Governing Transboundary Waters
Canada, the United States and Indigenous Communities
Emma S. Norman

Transboundary Water Politics in the Developing World
Naho Mirumachi

International Water Law and the Quest for Common Security
Bjørn-Oliver Magsig

Water, Power and Identity
The Cultural Politics of Water in the Andes
Rutgerd Boelens

Water and Cities in Latin America
Challenges for Sustainable Development
*Edited by Ismael Aguilar-Barajas, Jürgen Mahlknecht, Jonathan Kaledin
and Marianne Kjellén*

Catchment and River Basin Management
Integrating Science and Governance
*Edited by Laurence Smith, Keith Porter, Kevin M. Hiscock, Mary Jane Porter
and David Benson*

**Transboundary Water Management and the
Climate Change Debate**
*Anton Earle, Ana Elisa Cascão, Stina Hansson, Anders Jägerskog, Ashok Swain
and Joakim Öjendal*

*For more information and to view forthcoming titles in this series, please visit the
Routledge website:* http://www.routledge.com/books/series/ECWRM/

Stockholm International Water Institute (SIWI)

SIWI is a policy institute working for a water-wise world. SIWI does independent research, generates knowledge, and provides expert analysis and advice on water issues to decision-makers and other agents of change. SIWI organises the World Water Week in Stockholm – the leading annual global meeting place on water and development issues – and hosts the Stockholm Water Prize, the Stockholm Junior Water Prize, and the Stockholm Industry Water Award.

Transboundary Water Management and the Climate Change Debate

Anton Earle, Ana Elisa Cascão,
Stina Hansson, Anders Jägerskog,
Ashok Swain, and Joakim Öjendal

Routledge
Taylor & Francis Group
LONDON AND NEW YORK

from Routledge

First published 2015
by Routledge
2 Park Square, Milton Park, Abingdon, Oxon OX14 4RN

and by Routledge
711 Third Avenue, New York, NY 10017

Routledge is an imprint of the Taylor & Francis Group, an informa business

© 2015 **Stockholm International Water Institute**

British Library Cataloguing-in-Publication Data
A catalogue record for this book is available from the British Library

Library of Congress Cataloging-in-Publication Data
Earle, Anton.
   Transboundary water management and the climate change debate /
Anton Earle, Ana Elisa Cascão, Stina Hansson, Anders Jägerskog, Ashok
Swain and Joakim Öjendal.
      pages cm. — (Earthscan studies in water resource management)
   Includes bibliographical references and index.
   1. Watershed management—International cooperation.   2. Water-
supply—Co-management.   3. Climatic changes—Government
policy.   I. Title.
   TC413.E37 2015
   333.91'62—dc23
   2014046463

ISBN: 978-0-415-62975-1 (hbk)
ISBN: 978-0-415-83515-2 (pbk)
ISBN: 978-0-203-09892-9 (ebk)

Typeset in Bembo
by Apex CoVantage, LLC

The designations employed and the presentation
of material throughout the publication do not imply
the expression of any opinion whatsoever on the
part of UNESCO concerning the legal status of any
country, territory, city or area or of its authorities,
or the delineation of its frontiers or boundaries.

The book intends to provide a contribution to the
UNESCO IHP-VIII programme.

United Nations
Educational, Scientific and
Cultural Organization

International
Hydrological
Programme

MIX
Paper from
responsible sources
www.fsc.org   FSC® C013604

Printed and bound by CPI Group (UK) Ltd, Croydon, CR0 4YY

Contents

About the authors

Anton Earle, a geographer with an academic background in environmental management, specialises in transboundary integrated water resource management, facilitating the interaction between governments, basin organisations, and other stakeholders in international river and lake basins. Mr. Earle is the Director of the Stockholm International Water Institute (SIWI) Africa Regional Centre based in Pretoria, South Africa.

Ana Elisa Cascão is a Programme Manager at SIWI's Capacity Building services and co-director of SIWI's international training programmes on transboundary water management for the Lake Victoria, Middle East, and Tigris-Euphrates. Dr. Cascão is a Political Scientist with expertise in hydropolitics in the Middle East and North Eastern African regions and has published widely on this topic.

Stina Hansson, PhD in Peace and Development Research, at the School of Global Studies, Göteborg University, Sweden. She is doing research on transboundary water management and water services provision at Göteborg University and SIWI.

Anders Jägerskog, PhD is Counsellor for Regional Water issues in the MENA region at the Embassy of Sweden in Amman, Jordan and Associate Professor of Peace and Development at Goteborg University. The views expressed by Jägerskog do not necessarily reflect the views of the Swedish International Development Agency (SIDA) or the Swedish Government. Dr. Jägerskog was formerly Head of the Transboundary Water Management Unit at SIWI where he also served as the Programme Manager of the UNDP Shared Waters Partnership (SWP). UNDP partners with people at all levels of society to help build nations that can withstand crisis, and drive and sustain the kind of growth that improves the quality of life for everyone. On the ground in more than 170 countries and territories, UNDP offers global perspective and local insight to help empower lives and build resilient nations.

Ashok Swain is a Professor at the Department of Peace and Conflict Research and at the Department of Earth Sciences of Uppsala University, Sweden. He has written extensively on emerging security challenges and international water sharing issues.

Joakim Öjendal is a professor in peace and development research at the School of Global Studies, Göteborg University, Sweden. For over two decades, he has been doing research on transboundary water management and has published widely in the field.

Preface

Despite three decades of warnings and declarations, the global water crisis remains one of the unresolved and huge challenges to humankind. In fact, the water crisis is of such a magnitude that it is growing into an issue of a common global concern. And this situation is bound to get worse. Besides the obvious poverty creation inherited in, and development impediments held up by water scarcity, recent research is pointing out that inequality in water access is turning into political repression, gendered violence, and identity conflicts.

This perspective highlights transboundary water management (TWM): nearly 50 per cent of the global fresh water is to be found in 276 transboundary river basins; in Africa, it reaches 90 per cent, making the continent's development efforts hostage to effective governance of the shared water. The path to enhanced river basin management consists of two interwoven braids: knowledge (which we lack) and politics (which we do not understand).

Whatever the knowledge and understanding we do have, the existing agreements/regimes over transboundary rivers are increasingly turning volatile because of increased outtake and rendered unworkable due to the perception that global climate change (GCC) is altering the basic parameters for water governance. Whether physically correct or not – irrespective of which future scenario we apply – perceptions of GCC are undermining existing agreements, or are even instrumentally used to undermine the current water regime. Decreased river runoff is blamed on high evaporation rates or shifting rainfall patterns, while glacier lake outbursts seriously threaten water storage projects. Whereas governance and accountability always is an important couplet, in the international arena, GCC plays right into the securitization of transboundary water resources feeding a situation of non-transparency and non-cooperation.

The increasing global water scarcity and the changes (both real and perceived) that global climate change (GCC) will have on the management of transboundary water constitutes a concrete impediment to development efforts. Changes in quantity and quality of water – or the mere expectation of such changes – may provoke increased tensions and conflict; and hamper regional integration and development efforts.

In this book, we analyse the problematique of good governance of transboundary water resources. The reason for this undertaking is that the critical development-need for enhancing knowledge on how to most effectively govern these resources in the face of increasing uncertainties, and in particular in relation to GCC. With the help of in-depth study of six major transboundary river basins in Asia, Africa, and the Middle East, the book investigates the inflexibility of many basin agreements, regimes, and management principles in general. With the help of compared and contrasted findings from strategically situated case studies, the book provides generalized conclusions of relevance to a range of transboundary basins globally.

This co-authored book is a collaborative effort in the true sense. Each of us has tried to bring in her or his individual expertise and understanding base to support this group project. We have worked hard to provide a focused and integrated understanding of the problem and, at the same time, provide detailed analysis of cases.

The book is the result of a long, collaborative, and mostly enjoyable journey for which there are many to thank. Most of them are colleagues at the Stockholm International Water Institute (SIWI), School of Global Studies at Gothenburg University, and Department of Peace and Conflict Research at Uppsala University. We are also grateful to Florian Krampe for his assistance in editing the manuscript. The research for the book has been partly financed by the Swedish Research Council, the Stockholm International Water Institute (SIWI), and the United Nations Development Programme (UNDP).

Finally, we thank Earthscan (Routledge) Publishers for invaluable support to the project. We are particularly grateful to Tim Hardwick and Ashley Wright for their unlimited patience and insightful guidance.

Anton Earle
Ana Elisa Cascão
Stina Hansson
Anders Jägerskog
Ashok Swain
Joakim Öjendal

Foreword

UNESCO is deeply committed to addressing the challenges associated with the impact of climate change on water resources. Changes in the hydrological regime are increasing in many parts of the world, leading to extreme hydrological events and tensions over water that crosses political boundaries. *Transboundary Water Management and the Climate Change Debate* highlights the role that transboundary water management institutions can play in rising to these challenges and also the need for these organizations to adapt their respective mandates accordingly. The book contributes to the understanding of the importance of multidisciplinary research and the linkages between transboundary water management, climate change, peace, and water security.

Over the last decade, the debates on transboundary water management and global climate change have increasingly overlapped. This is a positive development, as much of the anticipated climate change will be clearly linked with effects on water. More water, less water, water at the wrong time and in the wrong place: climate change is bound to affect transboundary water management and the relations between states. This book sets forth practical examples of how climate change can affect water flows in basins such as the Nile, Ganges, Niger, Jordan, Orange-Senqu, and the Mekong, without attempting to present an exhaustive assessment of these phenomena. The authors approach the issue from the perspective of political discourse, and focus their attention on how the climate change debate is used in addressing dialogue on transboundary waters.

I hope this book will be a valued contribution to the debate on water diplomacy and shared water resources management, themes that are essential to the mission of the UNESCO International Hydrological Programme (UNESCO-IHP). IHP has long been engaged in transboundary water management through its role in supporting research on water cooperation. UNESCO's pioneering work, through the International Shared Aquifers Resources Management (ISARM) initiative and the From Potential Conflict to Cooperation Potential (PCCP) programme, has contributed extensively to the body of knowledge on transboundary water management and diplomacy.

Water cooperation is a subject fully included in the programme of the Eighth Phase of the IHP (2014–2021), 'Water Security: Responses to Local Regional and Global Challenges'. UNESCO is supporting the organization of training programmes and the strengthening of research capacities in this field throughout its network of centres and chairs. The authors of this book are collaborators of the newly established UNESCO International Centre for Water Cooperation (ICWC). This centre, established in 2014 and hosted by the Stockholm International Water Institute (SIWI), will play a pivotal role in fostering research and sharing knowledge in this domain.

For researchers and practitioners alike, this book offers important insights on the inherently political nature of water and climate change.

Some 276 river basins cross the political boundaries of two or more countries and 608 transboundary aquifer systems have been inventoried by the UNESCO IHP, but very few are the focus of projects that aim to improve the understanding of the link between transboundary water management and climate change.

I sincerely hope that this publication will raise awareness of these sensitive and urgent debates.

Flavia Schlegel
Assistant Director-General
for Natural Sciences
UNESCO

1 Introduction

Transboundary water management between a rock and a hard place

Watching CNN in late January 2011, the same week this book project was born, the first four news headlines were weather related – torrential rains with devastating mudslides in Brazil; cruel blizzards raging in the central United States bringing a large part of the country to a halt; flooding of the millennium tormenting north-eastern Australia (an area the size of France submerged); an emerging epic starvation in the Horn of Africa resulting from seasonal absence of rains – and all presented as global climate change related. Water movements were driving these disasters in different ways, and emerging with such supremacy that even the most powerful countries were failing to manage these disasters, let alone govern them proactively.

This volume recognises three levels where climate change 'exists': first, as a physical fact where human activity considerably alters the global climate. This is established as far as natural science can take us to-date.[1] Second, it is important to differentiate between climate change and climate variability. Natural climate variability is the norm, and some regions of the Earth have experienced large swings in climate over millennia. This variability can refer to long-term variations in climate as well as short-term fluctuations – inter-seasonal or annual. Third, climate change 'exists' as an issue of popular discussion; it has erupted out of the august corridors of scientific and academic research institutions and become a public topic, inciting a range of views and responses. The discussion is prominent on television talk shows, political debates, social media, conventional news, and during street protest movements. In the news press, public policy documents, and in popular perception there is a general underlying assumption that global climate change is responsible for extreme weather events, and that it is here to stay (and grow). This implies that the discussion around climate change carries political significance, and that there will be interests at various levels that will *adapt to it, resist it, and/or subvert* it. It is this last understanding of global climate change as a publicly discussed issue, with political dimensions, that is the focus of this book.

Therefore, this book provides no opinion on whether climate change is indeed the direct cause of current and future weather-related catastrophes. However, this link has been discursively established globally and reinforced through politicians, government officials, the media, academics, and civil society groups. Hence, the phenomenon of climate change has broken out from the strict, natural, science-based models and has – with a formidable, discursive power – entered the politics of constructing the root cause of the problems that we see around us, and is interpreting the possible solutions to the problem as it is constructed (cf. Edkins & Pin-Fat 1999). We are at the beginning of a global discussion with strong political motivations that will have major implications in many contexts. The way that transboundary water management institutional frameworks respond to this discussion needs to be understood, which in essence is the goal of this book.

A key issue to consider is the quality of transboundary water management (TWM) institutional arrangements, moving beyond counting the number of agreements between states in a basin or the presence or absence of a basin management organisation (cf. De Stefano et al. 2012). At the international level, this management process is projected to become more complex due to the pernicious impacts of climate change (Drieschova et al. 2009; Gleick and Cooley 2009; Falkenmark & Jägerskog 2010). This is a real dilemma since the 'unstoppable object' of the climate change discussion meets the 'unmovable structure' of conservative TWM practices; in this process, something will have to yield. Research on this dilemma has just emerged (Beck & Bernauer 2011; Tir & Stinnett 2012; cf. Huntjens et al. 2008), although the role of River basin organisations is uncertain and neglected (Tir & Stinnett 2012), and we certainly need to learn a lot in a fairly short time in order to adapt institutionally to the challenges of the climate change (Huntjens et al. 2012). Hence, this book aims to assess the impact of the climate change *discussion* on TWM and the institutions tasked to solve these problems. Essentially, we ask, 'What are the governance responses to the climate change discussion?' These responses could be seen in TWM institutions (basin-wide or on a sub-basin level), from individual basin states, or even from local institutions within a basin state. Let us consolidate and elaborate these ambitions below.

Framing the problem

Two major issues emerge from the growing public debate around climate change and impacts on freshwater resources, presenting those responsible for water management with a difficult decision.

First, climate change is happening. The Stern Review, the Intergovernmental Panel on Climate Change (IPCC), and the United Nations Framework Convention on Climate Change (UNFCC) have confirmed this, and there is now a popular understanding that climate change is real. The climate-related processes around us, to which we did not previously pay any particular attention – such as receding glaciers, lengthening of grazing periods on high-altitude plateaus, and the thawing of permafrost areas – are now popularly deemed part of the larger package of climate-change impacts. When news of exceptionally heavy rainfall, higher-than-average temperatures, or lengthened droughts is reported, the public typically ascribes these unusual meteorological conditions to climate change. These outcomes of climate variability have always been part of the picture in one way or another, but are now being portrayed as something new. At times it looks like politicians and technicians in the countries and/or regions affected are using climate change discussions to take the responsibility out of their hands, to justify their inability to deal with the recurrent climate disasters.

Global climate fluctuates naturally through numerous phenomena, but climate can be modified through human activities. The UNFCCC states in Article 2: "Climate change" means a change of climate which is attributed directly or indirectly to human activity that alters the composition of the global atmosphere and which is in addition to natural climate variability observed over comparable time periods' (UNFCCC 1994). According to the IPCC, human impact on global climate since the beginning of industrialisation exceeds the natural processes of climate variability (Solomon et al. 2007). Good water management strategies have always sought to take into account natural climatic variability; the difference now is that there is evidence and a belief that the water management practises of the past will not be sufficiently robust to respond to the new dimension of climatic change (Lenton & Muller 2009).

Whether climate change is man-made or 'natural', or indeed globally relevant, may be debated for decades, with various actors taking up positions driven by their interests. Unsurprisingly, the representatives of major carbon emitters, such as oil companies and coal mines, seek to downplay the scale of the changes, as well as the contributions from human activities. Likewise, promoters of alternative energy, flood defence, and water efficiency technologies warn of the dire consequences of inaction. The discussion around climate change has started to form a political and sanctioned discourse of its own, finding constituencies, and interpretations as well as scenarios and action plans (e.g. Huntjens et al. 2012). The politicised nature of the climate change discussion segues into the existing debates around water management and specifically TWM, where actors' responses mirror the partisan responses found in other sectors, as the case studies in this book will show. As Castree (2004:163) has commented, 'Geographical imaginations matter. This is precisely why we need constantly to interrogate their presuppositions, as well as the kind of world they aim to engender'.

Climate change, due to greenhouse gas emissions, is predicted to particularly affect the variability of water resource availability globally. It is true that the IPCC has come under question – for instance, over the prediction that the Himalaya glaciers could possibly melt away as early as 2035. In a statement the IPCC confirmed that this prediction was wrong and not based on rigorous scientific modelling (IPCC 2010). Nonetheless, in the same statement the IPCC affirmed that its conclusions that 'climate change is expected to exacerbate current stresses on water resources from population growth and economic and land-use change, including urbanization' is still scientifically verified and remains unchallenged

(IPCC 2010). Moreover, the increase of global surface temperature, through the greenhouse effect, is expected to increase the 'atmospheric water-holding capacity' (Min et al. 2011), among others, because of increased evaporation. That means scientists expect changing patterns of precipitation 'in the intensity, frequency, and duration of events' (Trenberth et al. 2003 in Yilmaz et al. 2014). As a consequence, floods and droughts will become more frequent. It is difficult to attribute the increased number and intensity of droughts and floods to climate change by excluding natural climate variability. However, recent studies seem to confirm that human-induced climate change already contributes to today's 'more-intense precipitation extremes' (Min et al. 2011) and can, for instance, be attributed to recent floods in the United Kingdom (Pall et al. 2011).

Scientific evidence is clear that the climate is changing (despite the magnitude and the impacts of the changes being under debate); but evidently – and more important for this volume – discursively, something is changing and growing in magnitude. The size, existence, and pace of the physical changes are primarily understood through complex global and regional circulation models that only highly trained scientists can access and interpret. In contrast, the politics of understanding and responding to climate change is open to a far wider audience; politicisation is the ultimate result. Echoing Buzan (1995), the climate change debate has become 'securitised' and has jumped the queue of political priorities, implying that climate change can have direct impacts to human, state, and even regional security. Climate change is presented as an existential threat to society, with direct impacts for human and state security. A 'security logic' naturally moves the issue higher up the political agenda, which is clearly what can be noticed with regards to climate change in recent years (Jägerskog 2008; Swain 2013).

A range of government and security entities have now responded to the possible threats posed by climate change. A report by the German Advisory Council on Global Change (WGBU) predicts that 'without resolute counteraction, climate change will overstretch many societies' adaptive capacities within the coming decades. This could result in destabilization and violence, jeopardizing national and international security to a new degree'. In the United States, the Pentagon, in its Quadrennial Defence Review of 2010, identifies climate change as one of the 'trends . . . whose complex interplay may spark or exacerbate future conflicts' (Department of Defence 2010), while the Central Intelligence Agency has established a Centre for the Study of Climate Change. The charter for the CIA's centre is 'not the science of climate change, but the national security impact of phenomena such as desertification, rising sea levels, population shifts, and heightened competition for natural resources' (CIA 2009). In NATO's 'Analysis and Recommendations of the Group of Experts on a New Strategic Concept', climate change is identified as one of the probable unconventional threats to the alliance and its partners (NATO 2010). A scoping study performed by the OSCE on the security implications of climate change for Europe concludes that 'Climate change will alter the socio-economic foundations of society. It will transform constants into variables . . . A particularly complex challenge is the water-food-energy nexus'. The study uses the term 'threat multiplier' to indicate that climate change may 'contribute to insecurities and the likelihood of armed conflict depending on given circumstances and the interaction with other factors' (OSCE 2010). The UNDP in its 2011 Human Development Report points out that 'past patterns suggest that, in the absence of reform, the links between economic growth and rising greenhouse gas emissions could jeopardize the extraordinary progress in the HDI in recent decades' (UNDP 2011). The message is clear – society needs to alter the way it responds to the threat of climate change impacts, as failing to do so will lead to environmental, economic, and security collapse. The majority of governments and international organisations have issued similar statements or policies, firmly placing climate change in the realm of security threats to which effective responses need to be developed (Jägerskog, Swain, & Ojendal 2014). The GCC discussion has also been adopted by the actors and institutions working in the field of TWM as a security threat, although the issue is being dealt with differently according to the specific hydropolitics of each of the transboundary river basins, or the regional settings of each particular river basin.

Furthermore, the climate change-related disasters observed above – of course, being mere illustrations of the wider problem – are primarily about water and water management. The exercise of good governance in transboundary contexts is increasingly caught between a 'rock' of conservative international water agreements on the one hand, and the 'hard place' of the ever-present discussions on climate change. And the need to deal with this situation is urgent. That there is a situation of crisis over the management of water resources of the world is a claim promoted by a range of international organisations (UNDP 2006; UNESCO 2008; UNEP 2009). Ecosystems are under increasing pressure from human activities and settlements; industries and cities have to convey water over increasing distances; and the production of crops for food and fuel uses ever-greater amounts of water – and yet, over a billion people globally lack decent access to water (UNDP 2006). This has led some researchers to conclude that 'for the first time in human history, human use and pollution of freshwater have reached a level where water scarcity will potentially limit food production, ecosystem function, and urban supply in the decades to come' (Jury & Vaux 2007).

The floods and droughts prevalent in many parts of the world have an impact on a variety of scales; leading to economic losses, stunted growth in children, movements of people to cities, as well as migration to wealthier parts of the world. Lacking are effective institutions for the sustainable development and management of the world's water resources. Coupled with the dynamic challenges posed by climatic change, there exists a real possibility of conflicts over water escalating at various levels. The UN Human Development Report of 2006, which focuses on water issues, concludes that the 'scarcity at the heart of the global water crisis is rooted in power, poverty and inequality, not in physical availability' (UNDP 2006, s. 2). In essence, this is a political problem. The 'Global Water Crisis' has been a crisis in its own right for at least two decades, emerging partly due to population growth, modernisation, and governance failure on several levels in the global system (Biswas 2009; Gleick 2009). While the extreme climatic events may not be possible to govern in a structured way (by definition being outside of the bounds of standard management approaches), the incremental change to local and regional climate is possible to plan and prepare for. However, it is a complicated enough task to manage and allocate transboundary water resources in 'normal' circumstances, and this is further complicated by the added uncertainty introduced by climate change. In general, the observation is that regional agreements and institutions, even when they are in place and are more or less effective in fulfilling their mission, are not yet prepared to deal with the issue of increasing climate change patterns (Tir & Stinnet 2012).

This is where the two points made above coincide. That is, the debate around climate change being taken up by the public at large and becoming politicised – even securitised – intersects with the challenges posed in managing water resources at the transboundary scale. Both are intense challenges in their own right, and they are getting increasingly connected. For political tensions to increase, it does not necessarily matter if climate change impacts hydrological conditions tangibly, as long as it is commonly *understood* to change them (or even argued to change them, politicising the issue). Either way, negotiation space is opened, and previously marginalised stakeholders (including state and non-state actors) will seek to push their positions forward, using the window of opportunity offered by the climate change debate. From a political point of view already unclear, unknown or contested knowledge on water availability is further brought into question; this challenges existing legal agreements and potentially complicates ongoing negotiations. Hence – in contrast to how the water management sector used to work (Falkenmark & Jägerskog 2010) – *the future* is added as an uncertainty factor, making it increasingly difficult to close (or uphold) any long-term agreements, including exact figures on water allocation. It is unusual for climate variability to be incorporated in the terms of agreements, though this has been observed in some recent agreements on transboundary river basins, such as the Incomati River in Southern Africa (Ashton et al. 2006).

In this volume the aim is to increase the understanding of perceived climate change impacts – and the challenges that they bring – in the context of international TWM. A particular focus will be on the responses (discourse and action) of the TWM institutions (including organisations, laws, and cultural practises) in the face of the climate change discussion. This will be done through developing an analytical framework of responses at various levels and exploring the empirical findings in six case-basins. The immediate and tangible actions which need to be taken are on the adaptation side, building resilience to the perceived impacts of climate change at local, national, and basin level. The mitigation side of the discussions will be included only where such actions are explicitly invoked by actors involved in TWM institutions.

Hence, climate change is rapidly emerging as a critical issue in the sharing of international waters and their attendant negotiation processes. In the past, water-sharing matters could be covered effectively by a few negotiators trained specifically to deal with water resource issues. But today, negotiators on transboundary watercourses need to have an understanding of an increasing range of fields (such as energy generation, food production, human rights, and health issues) and have sufficient knowledge and political understanding of possible impacts of climate change (such as precipitation patterns, glacier melting, temperature increase, and rising sea water encroaching on fresh water systems). Many riparian countries not only have to survive with the existing power asymmetry in relation to 'hydro-hegemons' in the basins, they also suffer from a lack of competent 'hydro-diplomats' who can address climate change issues while carrying out negotiations over TWM (Cascão & Zeitoun 2010). However, one would expect that the climate change discussion has the potential to be used by the different riparian states to serve their interests in the dialogue and negotiations over transboundary water resources. Responses to climate change discussions can eventually be used both as a hegemonic tool, as well as a counter-hegemonic tool with the potential to level the playing field of the less-powerful riparian states in a particular transboundary river basin.

It is likely that hydro-diplomacy – that is, the negotiations of underlying interests and issues in international water contexts (WWC 2012) – will start to take seriously the idea that the very basic parameters are changing through the climate change discussion, which will add a layer of complexity to future negotiations and agreements, and a layer of politicisation to the already concluded ones. Transboundary water negotiators will benefit from having a deeper knowledge of the climate change discussion and the possible impact of climate change on social, economic, and environmental factors in countries and regions. There may also be a demand for knowledge on the existing and emerging schools of thought regarding climate change and its impact on water availability and demand. It is crucial to identify and classify important actors and groupings, and their positions on climate change and water management issues. Moreover, hydro-diplomacy must have an overview of legal and policy documents (which are produced by international and regional organisations) on the impact of climate change on water resources, and possible mitigation and adaption measures.

The climate change discussion has entered the political realm where it matters in small and large decisions, adding to the existing global water crisis; presenting further critical elements to a *problematique* already suffering from complexity, politicisation, and securitisation. Or put differently, we have come to a point where we need to realise that there is *pressure for change* to improve global water management, and that this needs to take place on global, regional, and national levels. This volume aims to make a contribution to reaching a better understanding of these processes and how they are responded to, particularly at basin level.

Framing the institutional response to discussions on global climate change

To the degree that international water is regulated, it is typically done through an institutional framework; comprised of bilateral and/or multilateral organisations and bilateral and/or multilateral agreements (cf. Wolf 1995; cf. Conca

2006; cf. Earle et al. 2010). Many of these share some characteristics: they are typically the result of drawn-out bilateral dialogue and/or international multi-party negotiations over decades or more; cooperation is managed by a river basin organisation (RBO) of some sort (cf. Swain 2004); agreements are sometimes defined in figures, for example in terms of water allocation (water quantity or percentages), seasonal calendars, and long-term data on average rainfall/water-flow patterns (Jägerskog & Falkenmark 2010); and they are based on a political understanding emanating from a watercourse-related base (such as 'prior use' or 'equitable share'). As a result, it seems, water agreements have a conservative nature – it takes a long time to negotiate them, and once negotiated, they tend to stay in place for a long time. It is far from certain that parties to a water agreement which includes substantive provisions such as defined water availability, exact allocation formulas, and distinct minimum flow volumes are keen to renegotiate these agreements. Especially not if they think they have a good deal as it is and have invested heavily in the situation emanating from that deal, even if massive evidence on changed water regime surfaces from the climate change discussion. On the other hand, those with a (perceived) less favourable deal may be pursuing the necessity to re-negotiate based on climatic changes – even when these impacts on the water resource are not well established.

Climate change has added increased uncertainties to the smooth functioning, even survival, of international water agreements. As Arnell argued in 1999, 'climate change may affect the demand side of the balance as well as the supply side' (p. S32). With increasing temperatures, there is a possible sizeable reduction in precipitation in some basins, resulting in reduced water supplies available to users. Climate change will not only decrease the supply of usable water in many parts of the world, it may also increase water demand in domestic, agricultural, industrial, and ecological settings – driven by higher rates of evaporation (REF). Thus, climate change-induced scarcity and uncertainty around the quantity and quality of shared water resources in the arid and semi-arid regions can possibly increase competition for water resources amongst countries. In addition to the projected impacts on the quantity of precipitation a basin receives, it is likely that the timing will also be affected. Increased variability of precipitation will see greater inter- as well as intra-seasonal changes, with flash floods and dry spells becoming more common. This further enhances the uncertainties and anxieties over the water availability in the shared river systems. As Drieschova and others argue, 'variability of water flows can create risks for the longevity of agreements, because it is a change of circumstances which may cause states to change preferences, thereby reducing incentives to follow agreements signed in the past' (Drieschova et al. 2009: 292). Most of the existing river agreements do have provisions to meet short-term short falls in the flow volume (Tarlock 1999–2000). However, climate change can potentially result in long-term changes to water availability, which requires water regimes and institutions to be flexible and robust enough to adapt to a changed situation (McCaffrey 2003; Swain 2012).

Agreements and institutions at the regional level

The signing of agreements on several large or locally significant river and lake basins like the Ganges, Mahakali, Jordan, Lake Victoria, Mekong, Nile, Orange-Senqu, Zambezi, and Limpopo in the last two decades highlights the role water can eventually play in promoting peace and cooperation between states. However, the majority of the international basin agreements do not include all the states in a basin; and water management regimes and institutions are often dictated by the most powerful riparian(s). In all these basins, it can be argued that little progress has been made in establishing water management institutions which deliver effectively on their respective mandates. In Table 1.1 we provide an overview of the various TWM institutions to be found in basins – interpreting the term 'institution' in its broadest sense to include organisations, laws, and cultural norms. In essence, neither of the basin institutions typically delivered fully on their respective mandates, nor did they contribute tangibly to socio-economic development or ecosystem protection objectives, as defined in the missions of these institutions (Cleaver & Franks 2005; Molle 2008). Regional and national geopolitics are usually obstacles for a transition from bilateral sharing arrangements to multilateral basin-based development, although this varies from river basin to river basin. However, there is uncertainty over whether the present form and progress of the TWM agreements and institutions are suitable to meet challenges (such as climate change) that were not present when the agreements were negotiated and subsequently signed. In some cases, river basin multilateral initiatives and institutions have come about as a result of pressure from external partners, and they survive mainly with outside financial support. It is not yet clear what the future of these organisations will be. This exposes the differences in objectives for these organisations between the basin states and outside partners; where the former may be more concerned with stimulating sustainable socio-economic development, international partners are more concerned with environmental protection.

The transboundary institutional structures include river basin organisations (of various sorts), and laws and agreements, as well as norms and values held by the societies concerned. In this book, all forms of TWM organisations mentioned in Table 1.1 are referred to as river basin organisations (RBOs). We recognise that several of the organisational types do not conform entirely to a strict definition of an RBO (for instance, by not possessing a legal personality), but for convenience, the term RBO is used broadly. Likewise, the various legal instruments entered into between states are variously referred to as laws, agreements, conventions, or memoranda of understanding – and all being used synonymously for the purposes of this book. Generally, the term 'agreement' will be used by the authors – but at times, some of the other terms are used.

A wide range of actors are involved in the TWM institutional frameworks, some explicitly while others implicitly. Traditionally, the responsibility for the management of transboundary waters has been vested with the national government institutions, with very little delegation to other groups (Earle et al.

Table 1.1 Categories of TWM intuitional arrangements considered in this study

Category	Sub-Category	Scope	Specific issues	Nature
legal	Water Agreements	Specific legal agreements for water issues	Water quantity/allocations; Water quality; Environmental protection; Navigation	Bilateral/Multilateral
	General Agreements	Peace or cooperation agreements	General agreements but with specific articles on water resources management	Bilateral/Multilateral (but usually more of bilateral nature)
	Regional Water Conventions	Regional conventions on water resources	Principles on water management, environmental protection, cooperation, conflict resolutions mechanisms, etc., applied to the region	Regional (countries can sign and ratify, and adopt it at national level)
	General/Global Framework Conventions	Global conventions on water resources, or other water-related issues	General legal principles, such as equitable utilisation, no–harm, obligation to cooperation, etc.	Global (countries can sign and ratify, and adopt it at national level)
	Memorandum of Understanding; Protocols; Accords; Exchange of Letters	Legal instruments to address specific issues (not legally binding)		Bilateral/Multilateral
Organisational	Joint Water Committees	Dealing with specific water issues	Water quantity/allocations; Water quality; Monitoring role; Usually very narrow scope, without any development of infrastructure role	Bilateral/Multilateral
	River Basin Initiatives	River basin organisations without legal personality	Can cover several technical, development or infrastructure issues (but only advisory or consultation roles, because of lack of legal persona)	Multilateral (not necessarily including all riparian states)
	Development Authorities	River basin organisations with permanent legal personality	Water quantity/allocations; Water quality; Regional socio–economic development (agriculture, energy, watershed management, etc); Infrastructure development	Bilateral/Multilateral
	River Basin Commissions	River basin organisations with permanent legal status	Advise member states on the management and development of basin; Usually not with strong executive powers; Regional Socio-economic development (agriculture, energy, watershed management, etc.)	Multilateral (not necessarily including all riparian states)

2010; Earle forthcoming). Issues of sovereignty and state security have precluded the type of devolution of responsibility for water resource management experienced at the sub-national level, where catchment agencies or water user groups could assume responsibilities. Legal negotiations between countries are conducted primarily by state representatives (usually including national departments responsible for water or the environment, as well as foreign affairs and possibly the judiciary), with occasional input from expert groups on some legal or technical issues. Likewise, the majority of basin-wide organisations (be it joint water committees, river basin initiatives, development authorities, or river basin organisations) formed as a result of interstate negotiations, are either comprised exclusively of or dominated by government representatives. Often, professionals employed by these organisations are answerable directly or indirectly to their respective government representatives. In some cases other stakeholders are invited to make inputs to the meetings of the basin organisation either on an ad-hoc basis or through a standing arrangement. An example of the latter is in the Permanent Okavango Commission (OKACOM), where a grouping of stakeholders (the Basin Wide Forum) is included in regular Commission meetings; allowed to raise items on the agenda and participate in discussions, however they do not have a voting right (Earle & Malzbender 2007).

Stakeholders in transboundary water management

Inter-state agreements and institutions at basin level are negotiated, signed, ratified, and established by state-level stakeholders, the *explicit stakeholders* in TWM. But ultimately the agreements and institutions will be implemented by stakeholders at sub-national level. In practise, the agreements negotiated at the international level by the respective governments are implemented at the national and local level, thus involving directly a range of other stakeholders. Local governments (such as municipalities), businesses, farms, and community groups may find their water-use activities curtailed (or increased) due to the provisions of an international agreement. As international agreements which a state has committed to take precedence over domestic law, it falls on the various local water management structures to bring these activities in line with the commitments made at the international level (Malzbender & Earle 2007). These are the *implicit stakeholders* in the TWM institutional framework, who exercise their power and influence through being able to appeal to the political powers at a national level, where ultimate responsibility for issues lie (Earle et al. 2010). Issues identified as a threat at the local level, whether the perceived impacts of climate change or the perceived impacts of climate change mitigation or adaptation measures (for instance, cut-backs in carbon emissions or changes to land use practise to increase floodplains) can rapidly escalate to higher political levels. In this way the implicit actors in TWM may have an influence at the international level, especially around an issue as charged with emotive response as climate change.

Metaphorically speaking, it is useful to understand the TWM institutional framework not as a monolithic unit, but rather as a series of cogs all exerting different forces, pressures, and inertia, both exogenous and endogenous, on the riparian states sharing a watercourse. Some of the cogs exert more power than others, such as the politicians at national level, while some may be wishing to turn in the opposite direction to the others. Other cogs in this system include the water management community (those who use and manage the water resource directly), as well as outsiders such as the researchers and international development partners (Earle, Jägerskog & Öjendal 2010). Overall, the system progresses in a series of fits and starts. The analysis of this book will seek to understand better the institutional response of the broader framework around TWM, not as a single unit but rather comprised of a range of actors – implicit and explicit. The proposal is made that TWM institutional response to the climate change discussion falls in three broad categories, with some possible overlap between them. The responses are to *adapt* to, to *resist,* or to *subvert* the climate change discussion.

While looking at the different categories of TWM arrangements, as well as the different stakeholders involved and influenced by these arrangements, it will be possible to analyse how in the last decade the climate change discussion has been adopted and/or adapted to serve the political or socio-economic interests of the several actors. When a certain debate, discussion, or concept is characterised by ambiguity, it can be used as well as manipulated by different actors with different objectives. In brief, different actors can:

  Adapt: by adopting climate change-oriented concepts, promote legal instruments, establishing institutional frameworks and implement action plans at the regional/basin, national, and even local level;
  Resist: by refusing or contesting the fundamentals of the climate change discussion, and deliberately disregard it in the policies and discourses adopted in the management of transboundary water resources;
  Subvert: by using the climate change discussions' concepts, legal, and institutional frameworks for other purposes rather than to adapt/mitigate the climate change impacts.

Let us further discuss these three concepts and their significance for the overall structure and the content of this volume.

Adapt to GCC discussion

For the purposes of this book, institutional *adaptation* to the climate change debate is defined as taking steps to promote the resilience or robustness of the institutional framework and how these 'allow' (and disallow) actors' interests and actions to shine through, in order to counter the current and potential impacts of climate change. Thus, one can say that there is adaptation when capacity is built to prepare and plan for actions, which need to be taken in response to climate change impacts on society, environment, and the political economy in general. These actions are at the *organisational level* where staff with the required skills are employed (such as climate change specialists or flood and drought forecasters) or when training on these technical issues is promoted; at the *systems level*, by setting in place early-warning systems or dam synchronisation mechanisms; or at the *structural level*, by increasing the storage capacity of dams or developing inter-basin transfers or building flood defences, in order to deal with climate change. The UNECE in its guidelines on adaptation to climate change for water institutions envisages actions along an *adaptation chain* (see Figure 1.1), consisting of: prevention, improving resilience, preparation, reaction, and recovery (UNECE 2009). The first two actions, prevention and improving resilience, are related both to the gradual effects of climate change and to extreme events. The last three actions – preparation, response, and recovery – relate to responses to extreme events (or their possibility) such as floods and droughts. At the transboundary level the range of actions will be considerably narrower than at the national level – with an expected emphasis on the prevention and resilience actions. But as the case studies will show, at the TWM level and in particular in the contexts where river basin organisations have been established, there is a growing trend for the inclusion of plans, policies, and projects addressing climate change issues. Whether these actions are a mere cosmetic operation for political or financial purposes will be discussed later in this book.

Resist the GCC discussion

Institutional *resistance* to the climate change discussion could be either passive or active. Passive resistance would amount to avoidance of the issue, essentially seeking to wish it away through ignoring it. That would mean that in activities of TWM institutions, such as basin planning, water allocation, infrastructure development, and data management, the climate change debate is not incorporated in any explicit way. Natural climatic variability may well play a role in the institutional activities, but at no time is there explicit reference to climate change. A more active form of resistance to the climate change debate would be to question the legitimacy of the forecast impacts of climate change – essentially participating directly in the debate about the validity of climate change.

The resistance to incorporating elements of the climate change debate in TWM institutions is most visible in two main areas. The *first* is in the investments made in existing infrastructure. If water has been allocated and diverted to supply a specific user group, it becomes difficult to re-direct it, even in the face of changes to the hydrological conditions. As Allan points out, any society which has allocated water to irrigation has eventually suffered water scarcity (Allan 2011). The political costs of taking water away from existing users, especially in the agricultural sector, are high. The result is that the allocation of water and the operation of infrastructure in such cases are unmodified by the climate change debate. The Aral Sea fits this mould, where water from the Amu Darya and the Syr Darya rivers was diverted upstream of the sea to cotton plantations in what is today Uzbekistan (Micklin 2007). Since the water diversion started in the 1960s, there was a 90 per cent decrease in the volume of the sea by 2007. In addition,

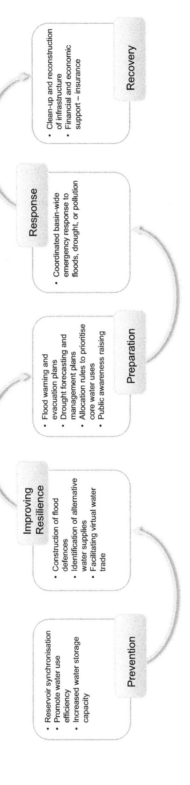

Figure 1.1 The climate change adaptation chain: institutional actions at the transboundary level (Redrawn from UNECE 2009)

the 'salinity level increased from 10 to more than 100 grams per litre, causing negative ecological changes, including decimation of native fish species, initiation of dust/salt storms, degradation of deltaic biotic communities, and climate change around the former shoreline' (Micklin 2007: 47). The negative impacts of the diversion of water have been exacerbated by global climate change, leading to higher rates of evaporation and greater environmental problems (Micklin 2007). Despite this situation it has not proved possible to reach an agreement whereby water is taken away from the cotton fields in Uzbekistan and reverted to the Aral Sea.

The *second* area where the resistance to incorporating the climate change debate is evident is around international agreements on water management. These have either been concluded prior to the issue of climate change coming to the fore or are in the process of being negotiated, but due to their complexity ignore changes to climate. The result is that agreements are not flexible enough to allow for climate change, thus becoming a possible point of friction between countries; something they were designed to reduce (Falkenmark & Jägerskog 2010). The reality is that the majority of agreements concluded on transboundary watercourses devote little attention to climatic variability and even less to global climate change (cf. Huntjens 2012). Sometimes agreements have also been formulated with a 'constructive ambiguity', thereby avoiding some of the challenging issues, including climate variability and climate change. This is primarily done in order for the respective parties to be able to, in their respective constituencies, 'sell' the agreement as something benefitting them. Ambiguity permits variations in interpretations that are useful to governments, but which also results in problems and challenges in implementation. This situation has been argued to be the case in the Jordan-Israeli Peace Agreement from 1994 as it relates to water (Jägerskog 2003).

No matter whether resistance to the climate change debate is passive or active, the result is that the issue is not incorporated in the mandated activities of TWM institutions. Given the current politicised nature of the climate change debate and the popular opinion it evokes, it is unlikely that institutional actors would totally ignore it. Thus the act of omission (from institutional activities) becomes an action in itself; something of which to take note.

Subvert the GCC discussion

In transboundary river basins we can find examples of actors and institutions that adapt to and resist the climate change discussion, as mentioned above. At times the climate change discussion is also subverted; that is, the initial meaning of the discussion is transformed in order to suit the interests of certain actors. This becomes possible due to the uncertainty and ambiguity of the climate change studies and the ensuing debate around it. Playing with climate change ambiguity has been transformed into a tool for riparians and river basin organisations to influence the agenda-setting and investments in these basins. Interesting examples of how the climate change discussion is subverted at the basin level are: 1) new climate change-oriented resource mobilisation strategies; and 2) new discourse surrounding the construction of large-scale water infrastructure.

Since 2008, considerable funding for climate change strategies (both mitigation and adaptation) has become available internationally. In some cases, traditional funding for water resource management has been progressively transferred to climate change issues. Countries can now apply for funding that was not available before (in the case of money newly-committed to climate change issues), or for funds previously available for water resources management but now devoted to climate change issues by including climate change adaptation and/or mitigation strategies in projects. One of the aims of this book is to understand how climate change funding mechanisms impact both the action and discourse in TWM institutional frameworks. For example, what is the impact on the resource mobilisation strategies of RBOs and how are RBOs now packing their investment projects in order to attract funding?

Another development is related to the funding of large-scale water infrastructure. In the past two decades, and in particular since the World Commission for Dams Report (2000), funding for the construction of large-scale dams by traditional bilateral development partners,[2] multilateral development partners, and international and regional banks has been scarce. Most of the dams that have been built in the past decade have been financed by the countries building them and in collaboration with non-traditional development partners, such as Gulf countries, Islamic banks, China, construction companies, and commercial banks. However, the climate change discussion provides a new opportunity for traditional donors and multilateral development partners to get involved in the dam-construction business, in particular in Africa. The reason for this is that, according to the climate change discussion, large-scale dams can play a role in both climate change adaptation strategies, as well as mitigation strategies. The argument is that dams and their associated reservoirs have the potential for both flood prevention and control, as well as mitigation of drought impacts, increasing the capacity of countries and regions to adapt to climate change. Meanwhile, it is argued, hydropower is a reliable provider of 'clean energy' – contributing to efforts to mitigate the advance of climate change through the reduction of greenhouse gas emissions.

Institutional *subversion* of the climate change debate implies the evocation and manipulation of selected elements of the debate to serve the needs of a particular actor in the TWM institutional framework. Thus, an argument made in the debate around climate change is used to further the ends of an actor, serving as a justification of their position and interests. In such a case the potential impact of climate change is invoked to justify or block the development of water storage or transfer infrastructure. For instance, in October 2011 the Executive Secretary of the Lake Chad Basin Commission, Mr. Sanusi Abdullahi, attributed the shrinkage of that lake to the impacts of climate change. He stated that 'the current size of the basin is 2,500 square kilometres from its original size of 250,000 square kilometres in the 1960s; climate change has caused the shrinkage. As a result of shortage of rainfall over the years, many rivers including River Hadejia and Nguru in Nigeria have been affected with short supply of needed water' (NAN 2011). As a result there should be a transfer of water from the Congo River to the Lake Chad basin, argued the Executive Secretary. At the time he was making a direct appeal to the African Development Bank to provide funding for the project, adding that the restoration of the lake would provide social and economic benefits to all the states concerned and provide relief from the impacts of climate change (NAN 2011). Whether climate change is indeed the reason for the demise of Lake Chad is of less interest for this book than the fact that an official government representative invokes it to motivate for the support of a specific infrastructure project.

Adapt, resist, or subvert? Or all?

Overall, there is a growing and politicised discussion on climate change increasingly making its mark politically, socially, and economically on attempts to manage our physical environment; and we have a global water crisis with efforts to manage competing inter-state water needs at the basin level through TWM. The latter is volatile and imperfect (cf. Earle et al. 2010), but most importantly, it is to a large extent resting on the very foundation the discussion on climate change is altering – a predictable hydrological system. At this point, we do not know the depth or nature of the alterations precipitated by the climate change discussion, but we aim to find out through analysing a collection of cases in this book.

What is possible to conclude already at this point is that the climate change discussion is increasingly part of the TWM institutional arrangements, including agreements and regional organisations at the basin level. Examples of adaptation, resistance, and subversion of the climate change discussion can be found in several of the cooperation and conflict processes concerning the management of transboundary water bodies. Sometimes riparian countries assume different positions concerning the climate change discussion, for the purpose of their own national interests. But examples can also be found of riparian countries coming together to use the climate change discussion as a political tool in TWM processes, or when physical climate change impacts appear as a unifier and motive for cooperation between riparian states of the same basin. Examples of the first can be the new cooperative processes that include a component of fund-raising specific for targeting climate change-funding mechanisms. An example of the latter is when countries have started or have strengthened basin-wide cooperation after 'emblematic events', such as floods and droughts with severe regional impacts, and that acted as catalysts for collective action between riparian countries.

Power relation analysis is extremely relevant for the purposes of this book, as they are determinant for the TWM arrangements to be discussed further. Asymmetric power relations are usually at the core of negotiations and agreements between riparian states and the multilateral institutions establishment by them (Warner & Zeitoun 2008). How the climate change discussion is or is not integrated in these arrangements is also subject to asymmetric power relations. Due

to the current ambiguity associated with the climate change discussion, it is possible to assume that it can be used either way by the hegemonic actors – those who
have a stronger position in the basin – or by the non-hegemonic actors, who
have weaker access to and control of the transboundary water resources. By
adapting, resisting, or subverting the climate change discussion, different actors
can use it either to reinforce or to undermine the current *status quo* in the basin.
In any analytical framework that includes power analysis, it must be considered
that power structures cannot be taken for granted. Power and hegemonic structures do not remain unchanged or unchallenged over time (Cox 1983; Gill &
Law 1988). *Change* and *resistance* are also part of the dynamic process of political relations. Critical international relations give particular importance to the
nature of change in the processes of policy-making. Gill (1993) proposes that
any analysis requires the assessment of three interlinked issues: (a) the *nature* of
the political reality; (b) the key *components and relationships;* and (c) how these
change over time. We propose that the discussion around GCC and the institutional responses described above will impact on these three interlinked issues.

Although the analysis will focus mainly on the discussion around climate
change and its impacts on TWM institutional arrangements, the book does not
exclude the fact that extreme meteorological and hydrological events can be
relevant in the evolution of TWM arrangements in transboundary river basins.
As mentioned above, emblematic events, such as floods and droughts with severe
impacts on society, the economy and politics at regional level, can influence why
and how TWM arrangements are initiated, strengthened, or transformed. Faced
with climatic extreme events, actors might be more prompted and have more
incentives for collective action. Taking drought as an example: long-term droughts
are expected to reduce the water availability in the basin, with several countries
facing periodic water shortages. In the absence of infrastructure, technical knowledge and instruments as well as institutional frameworks that allow the countries to
cope with droughts, it is likely that the impacts will be severe. A reflexive process
by the riparian states and other actors at the regional and local level might act as a
trigger for cooperation, in order to prevent a repeat of severe impacts in the future.
In the Nile basin, for example, it is interesting to observe that after drought in the
upstream countries in the mid-1970s and the late 1980s, cooperation jumped the
queue of political priority. In the 1970s there originated a growing interest for
developing a regional platform for the collection and dissemination of hydromet-
ric and hydrologic data. In the late 1980s, after one of the worst droughts in the
Horn of Africa region, resulting in the lowest recorded flows of the Nile River,
the countries came together to find a common solution for the common problem.

We propose that the climate change discussion influences TWM institutional
arrangements, with responses falling into one or a combination of *adapt, resist,*
or *subvert.* These responses are, in turn, mediated by and amplified by the existing dynamic power relations between countries in a transboundary basin. The
institutional responses become evident either as actions or as discourse by the
actors involved. Based on the above, we work with six river basin cases, namely:
the Ganges-Brahmaputra, the Jordan, the Mekong, the Niger, the Nile, and the

Orange-Senqu. These river basins, although vastly different in size, flow, and political-economy, share certain key features:

- They are regionally important river basins in complex and politically contested contexts, comprising several riparians;
- They have all developed cooperative institutional frameworks (more or less successful in their ambition and implementation) and the basin states engage in dialogue on collaborative water management;
- They are all subject to intensive planning for the utilisation of available resources – either through a unilateral or a bi- or multi-lateral process;
- They are all situated in either water scarce areas or in areas highly vulnerable to change in climate and thereby water flows;
- Each basin has a clearly-identified hegemonic state – able to influence the pace and direction of collaborative management in the basin.

Together these cases will aid in exploring the analytical framework so far established, allowing the theoretical concepts to be enriched through empirical observation. This introductory chapter is followed by the six cases, each analysing the actions and discourse of TWM institutions in their respective basins. The book's concluding chapter draws together the lessons learned from the cases and re-interprets the analytical framework established in this chapter in light of the empirical findings.

In the case study chapters, after the stocktaking of the degree of vulnerability of the river basin to the possible impacts of the global climate change, there will be a close examination of capacity of the existing TWM agreements, institutions, and riparians about their coping abilities. Firstly, most of these existing arrangements provide some mechanisms to adjust to the runoff variability in short-term while agreeing on allocation of fixed quota of water to the competing riparian states. As global climate change assumes long-term changes to both demand and supply of river water, it is thus crucial to examine the suitability of existing agreements and institutions to address this challenge. Secondly, global climate change reports also projects the increase in the frequency of extreme events like droughts and floods in these basins, and it is also important to examine the ability of the water sharing arrangement arrived in 1990s to handle these disaster risks and uncertainties. At the same time, thirdly, there will be also an investigation of if there are new developments in the administrative structure, knowledge capacity, and legal character of theses TWM arrangements in the face of growing debates over the climate change related risks.

Some of the TWM institutions have originated with active involvement of international donor agencies and still surviving with their help and assistance. In others, some riparian countries are experiencing fast economic growth and gradually their demand for water resources is shifting from highly consumptive agricultural sector to pollution-prone industrial and urban sectors moving beyond a narrow development context and donor dynamics. In this context, it is crucial to investigate the state of the external actors' interest and incentive to invest for better TWM in the case basins. Moreover, the changing power balance in the basin due to the rapid rise of some basin states in economic and political terms also has the potential to affect the TWM institutions and their growth in both ways. Thus, there is a need for an assessment of increased or decreased internal investment within the basin by particularly the key riparian countries. Therefore, it is important to examine the strength and weakness of regional bodies and their ability to contribute towards better adaptation of TWM institutions to the climate change debate.

Finally, in addition to existing regimes, institutions, and RBOs, all the case study basins do have their own hydro-hegemons. These hegemons may avoid/refuse to compromise their dominant position and avoid sharing the water in an equitable manner with other weaker riparian countries. Thus, it will be important to investigate whether the change predictions regarding climate change impacts will have the ability to force the hegemons to adopt a problem-solving strategy in the basin, paving the way for the formation of effective basin-based TWM institutions for GCC. Or, if the perceived insecurity and enhanced rivalry will harden their positions further and leading to greater conflict level and possibly lower cooperation (and efficiency).

In the concluding analysis of the book, there will be an exhaustive assessment of the coping abilities of existing TWM institutions in individual cases and also analysis of how they stand in a comparative perspective. Each key factor will be assessed and evaluated. To measure the effectiveness of the TWM institutions, a detailed and careful evaluation will be done, taking social, economic, and political factors into account. Pursing a unified and organised structure for analysing and evaluating our case studies will facilitate making some generalised findings through the help of systematic and careful comparison.

Notes

1 The UN Framework Convention on Climate Change (UNFCCC) states that 'Climate change means a change of climate which is attributed directly or indirectly to human activity that alters the composition of the global atmosphere and which is in addition to natural climate variability observed over comparable time periods'. Climate change, driven by human activity, is indeed taking place, and water is at the core of the impacts of climate change on the hydrologic cycle, with an increasing convergence of the results of the various predictive models. UNFCCC (1992), United Nations Framework Convention on Climate Change, New York: United Nations. IPCC (2007). A body of scientific literature now exists on the likely impacts.
2 Note that traditional development partners in this book are defined as the 24 countries which in 2012 are members of the Organisation for Economic Cooperation and Development's (OECD) Development Assistance Committee (DAC).

References

Allan, Tony, 2011, *Virtual Water: Tackling the Threat to Our Planet's Most Precious Resource*. London: I. B. Tauris.
Arnell, N. W., 1999, 'Climate change and global water resources'. *Global Environmental Change*, 9, S31–S49.

Ashton, P. J., Earle, A., Malzbender, D., Moloi, M.B.H., Patrick, M. J., and Turton, A. R., 2006, *A Compilation of All the International Freshwater Agreements Entered into by South Africa with Other States.* WRC Report No. 1515/1/06. Pretoria, Water Research Commission.

Beck, Lucas, and Thomas Bernauer, 2011, 'How will combined changes in water demand and climate affect water availability in the Zambezi river basin?', *Global Environmental Change*, 21 (2011): 1061–1072.

Biswas, Asit K., 2009, 'Water management: some personal reflections', *Water International*, 34(4): 402–408.

Buzan, Barry, 1995, 'The levels of analysis problem in international relations reconsidered', *International Relations Theory Today.* Cambridge, UK: Polity Press.

Cascão, A. E., and Zeitoun, M., 2010. 'Power, hegemony and critical hydropolitics', in Earle, A., Jägerskog, A., and Öjendal, J. (eds.), *Transboundary Water Management – Principles and Practice.* London: Earthscan: 27–42.

Castree, N., 2004, 'Differential geographies: place, indigenous rights and "local" resources', *Political Geography*, 23(2): 133–167.

CIA, 2009, 'CIA opens Center on Climate Change and National Security', accessed 11 February 2011 at https://www.cia.gov/news-information/press-releases-statements/center-on-climate-change-and-national-security.html

Cleaver, Frances, and Tom Franks, 2005, *How Institutions Elude Design: River Basin Management and Sustainable Livelihoods.* BCID research paper No.12, Bradford Centre for International Development University of Bradford.

Conca, K., 2006, *Governing Water: Contentious Transnational Politics And Global Institution Building.* Cambridge, MA: MIT Press.

Cox, R. W., (1983, 'Gramsci, hegemony and international relations: an essay in method', *Millennium-Journal of International Studies*, 12(2): 162–175.

Department of Defence, 2010, *Quadrennial Defense Review Report.* Washington, DC: DoD.

De Stefano, Lucia, Duncan, James, Dinar, Shlomi, Stahl, Kerstin, Strzepek, Kenneth M., and Wolf, Aaron T., 2012, 'Climate change and the institutional resilience of international river basins', *Journal of Peace Research*, 49: 193. doi: 10.1177/0022343311427416.

Drieschova, A., Giordano, M., and Fishlander, I., 2009, 'Climate change, international cooperation and adaptation in transboundary water management', in W. N. Adger, I. Lorenzoni, and O'Brien, K.'s *Adapting to Climate Change: Thresholds, Values, Governance.* Cambridge University Press.

Earle, A., forthcoming, *Transboundary water management beyond the state*, PhD dissertation, SGS, Göteborg University.

Earle, A., and Malzbender, D., 2007, 'Water and the peaceful, sustainable development of the SADC region', paper produced for the Safer Africa project *Towards a Continental Common Position on the governance of natural resources in Africa.* SaferAfrica, Pretoria.

Earle, A., Jägerskog, A., and Öjendal, J., 2010, *Transboundary Water Management – Principles and Practice.* London: Earthscan.

Edkins, Jenny, and Pin-Fat, Véronique, 1999, 'Introduction: the subject of the political', in Edkins, Jenny, Persram, Nalini, and Pin-Fat, Véronique (eds.), *Sovereignty and Subjectivity.* London: Lynne Rienner: 1–18.

Falkenmark, M., and Jägerskog, A., 2010, 'Sustainability of transnational water agreements in the face of socioeconomic and environmental change,' in Earle, A., Jägerskog, A., and Öjendal, J. (eds.), *Transboundary Water Management – Principles and Practice.* London: Earthscan: 157–170.

Gill, S. (Ed.), 1993, *Gramsci, Historical Materialism and International Relations* (Vol. 26). Cambridge, UK: Cambridge University Press.

Gill, S. R., and Law, D., 1989, 'Global hegemony and the structural power of capital', *International Studies Quarterly*, 475–499.

Gleick, Peter H., 2009, 'Facing down the hydro-crisis', *World Policy Journal*, 26(4): 17–25.

Gleick, Peter H., and Cooley, H. S., 2009, 'Energy implications of bottled water', *Environmental Research Letters*, 4(1). doi:10.1088/1748-9326/4/1/014009.

Huntjens, P., Pahl-Wostl, C., Rihoux, B., and Flachner, Z., et al., 2008, *The Role of Adaptive and Integrated Water Management (AIWM) in Developing Climate Change Adaptation Strategies for Dealing with Floods or Droughts: A Formal Comparative Analysis of Eight Water Management Regimes in Europe, Asia, and Africa*. Osnabruck, Germany: Institute of Environmental Systems Research, University of Osnabruck.

Huntjens, Patrick, Lebel, Louis, Pahl-Wostl, Claudia, Camkin, Jeff, Schulze, Roland, and Kranz, Nicole, 2012, 'Institutional design propositions for the governance of adaptation to climate change in the water sector', in *Global Environmental Change* 22: 67–81.

IPCC, 2007: Climate Change 2007: The Physical Science Basis. Contribution of Working Group I to the Fourth Assessment Report of the Intergovernmental Panel on Climate Change [Solomon, S., D. Qin, M. Manning, Z. Chen, M. Marquis, K.B. Averyt, M.Tignor and H.L. Miller (eds.)]. Cambridge University Press, Cambridge, United Kingdom and New York, NY, USA.

IPCC, 2010, IPCC statement on the melting of Himalayan glaciers, accessed 11 February 2011 at http://www.ipcc.ch/pdf/presentations/himalaya-statement-20january2010.pdf

Jägerskog, A., 2008, 'New threats? Risk and securitisation theory on climate and water', in Brauch, Hans Günter, Oswald Spring, Ursula, Mesjasz, Czeslaw, Grin, John, Kameri-Mbote, Patricia, Chourou, Bechir, and Birkmann, Jörn (eds.), *Coping with Global Environmental Change, Disasters and Security – Threats, Challenges, Vulnerabilities and Risks*. Hexagon Series on Human and Environmental Security and Peace, vol. 5. Berlin: Springer-Verlag.

Jägerskog, A., 2003, 'Why states cooperate over shared water: the water negotiations in the Jordan River basin', PhD dissertation. Linköping Studies in Arts and Science: Linköping University.

Jägerskog, A., Swain, A., and Ojendal, J. (eds.), 2014, *Water Security*, vols. 1–4, SAGE, London.

Jury, W. A., and Vaux, H. J., 2007, 'The emerging global water crisis: managing scarcity and conflict between water users', *Advances in Agronomy*, 1–76.

Lenton, R. L., and Muller, M. (eds.), (2009), *Integrated Water Resources Management in Practice: Better Water Management for Development*. London: Earthscan.

Malzbender, D. B, and Earle, A., 2007, *The Impact and Implications of the Adoption of the 1997 UN Watercourse Convention for Countries in Southern Africa*. Washington, DC: WWF.

McCaffrey, S., 2003, 'Water disputes defined: characteristics and trends for resolving them', in International Bureau of the Permanent Court of Arbitration (ed.), *Resolution of International Water Disputes*. The Hague: Kluwer Law International, 49–90.

Micklin, P., 2007, 'The Aral Sea disaster', *Annual Review of Earth and Planetary Sciences*, 35: 47–72.

Min, S. K., Zhang, X., Zwiers, F. W., and Hegerl, G. C., 2011, 'Human contribution to more-intense precipitation extremes', *Nature*, 470(7334), 378–381.

Molle, Francois, 2008, 'Nirvana concepts, storylines and policy models: insights from the water sector', *Water Alternatives*, 1(1):131–156.

NAN, 2011, 'Lake Chad basin shrinks to 2,500 Sqm', *Executive Secretary*. News Agency of Nigeria, accessed 11 March 2011 at http://www.nanngronline.com/section/africa/lake-chad-basin-shrinks-to-2500-sqm-%E2%80%93-executive-secretary

NATO, 2010, 'NATO 2020: Assured security; dynamic engagement', accessed 11 February 2011 at http://www.nato.int/cps/en/natolive/official_texts_63654.htm?selectedLocale=en#p1

OSCE, 2010, 'Shifting bases, shifting perils: a scoping study on security implications of climate change in the OSCE region', accessed 11 February 2011 at http://www.osce.org/eea/78357

Pall, P., Aina, T., Stone, D. A., Stott, P. A., Nozawa, T., Hilberts, A. G., . . . and Allen, M. R., 2011, 'Anthropogenic greenhouse gas contribution to flood risk in England and Wales in autumn 2000', *Nature*, 470(7334), 382–385.

Solomon, S., Qin, D., Manning, M., Chen, Z., Marquis, M., Averyt, K. B., . . . and Miller, H. L., 2007, 'IPCC, 2007: summary for policymakers', in *Climate Change 2007: The Physical Science Basis. Contribution of Working Group I to the Fourth Assessment Report of the Intergovernmental Panel on Climate Change*. Cambridge, UK: Cambridge University Press.

Swain, Ashok, 2004, *Managing Water Conflict: Asia, Africa and the Middle East*. New York: Routledge.

Swain, Ashok, 2012, 'Global climate change and challenges for international river agreements', *International Journal on Sustainable Society*, 4(1, 2): 72–87.

Swain, Ashok, 2013, *Understanding Emerging Security Challenges: Threats and Opportunities*. London: Routledge.

Tir, Jaroslav, and Stinnett, Douglas M., 2012, 'Weathering climate change: can institutions mitigate international water conflict?' in *Journal of Peace Research* 49: 211 doi: 10.1177/0022343311427066.

UNDP, 2006, *Human Development Report 2006: Beyond Scarcity, Power, Poverty and the Global Water Crisis*. New York: UNDP.

UNECE, 2009, *Guidance on Water and Adaptation to Climate Change*. Geneva: UN Economic Commission for Europe.

UNEP, 2009, 'Water at a glance: the global crisis', accessed 2010 at http://www.unep.org/ourplanet/imgversn/141/glance.html

UNESCO, 2008, 'From potential conflict to cooperation potential', accessed 2011 at http://www.unesco.org/water/wwap/pccp

UNFCCC, 1994, Article 1, definitions: UNBIS Thesaurus, accessed 6 February 2015 at http://unfccc.int/files/documentation/text/html/list_search.php?what=&val=&valan=a&anf=0&id=10

Warner J. F., and Zeitoun, M., 2008, 'International relations theory and water do mix: a response to Furlong's troubled waters, hydro-hegemony and international relations', *Political Geography*, 27:802–810.

Wolf, Aaron, 1995, *Hydropolitics along the Jordan River: Scarce Water and Its Impact on the Arab–Israeli Conflict*. Tokyo: United Nations University Press.

Yilmaz, A. G., Safaet, H., Huang, F., and Perera, B. J.C., 2014, 'Time-varying character of storm intensity frequency and duration curves', *Australian Journal of Water Resources*, 18(1): 15.

WWC, 2012, *Hydro-diplomacy in Motion*. Marseille, France.

2 The Ganges–Brahmaputra River basin

Introduction

The Ganges-Brahmaputra is one of the largest river basins in the world. This 1,634,900 square kilometres of basin spreads over Bangladesh, Bhutan, India, Nepal, and China (Tibet). The Ganges and Brahmaputra rivers both have their source in the Himalayas, where glacier melt water is an important source for the headwaters of the river. A number of studies have claimed in recent years that thanks to climate change, the Himalayan glaciers are melting more rapidly, and that can further bring major variation to the runoff of this river system (Yao et al. 2012). The Ganges River originates on the southern slope of the Himalayan range, and on its way seven major tributaries augment its flow. Three of them – the Gandak, Karnali (Ghagara), and Kosi – run through Nepal, and they supply approximately 60 per cent of the Ganges flow. After leaving Himalayas, the Ganges flows in the south-easterly direction through India, then forms the boundary between India and Bangladesh for about 112 kilometres before entering into Bangladesh. The Brahmaputra River originates from near Lake Mansarovar, and after flowing in China (Tibet) in an easterly direction comes to the north-eastern part of India, and then to Bangladesh to merge with the river Ganges in the middle of Bangladesh. The main tributary to the system, the Meghna River, originates in the northeastern region of India and then flows into Bangladesh to join the combined flow of the Ganges and Brahmaputra. Of the total basin area of this massive river system, 62.93 per cent is in India, 7.39 per cent in Bangladesh, 2.58 per cent in Bhutan, 8.02 per cent in Nepal, and 19.08 per cent in China (Tibet) (Verghese & Iyer 1993: 44).

The combine flow of the Ganges-Brahmaputra River runs to the Bay of Bengal, where it forms the Ganges Delta. This delta is one of the largest in the world and has the highest population density of any delta. The world's higher population density is supported by this important river system, making it important to the survival of millions of people. The total basin area is approximately 1.66 million square kilometres, of which the Ganges River supports more than 1 million square kilometres and is densely populated with approximately 500 million people. The Ganges part of the basin is highly agriculture intensive, as 71 per cent of it is cultivated. The Ganges-Brahmaputra River system, which

Table 2.1 Basic information on Ganges-Brahmaputra basin

Ganges River	Geographical Location	India: 62.9%; China: 19.1%; Nepal: 8.0%; Bangladesh: 7.4%; Bhutan: 2.6%
	Length of the River	2,525 kilometres
	Drainage Basin	1,080,000 square kilometres
	Basin Mean Annual Flow	525.02 cubic kilometres
Brahmaputra River	Geographical Location	China: 50.5%; India: 33.6%; Bangladesh: 8.1%; Bhutan: 7.8%
	Length of the River	2,900 kilometres
	Drainage Basin	580,000 square kilometres
	Basin Mean Annual Flow	629.05 cubic kilometres

(Source: Singh & Arora 2007)

carries such a huge significance in the lives of millions of people in the South Asian region, suffers from high seasonal fluctuations in its flow.

Water availability in the basin is very uneven, both temporally and spatially. The Ganges River has a peak flow of 141,000 cubic metres and drains 1,150 cubic kilometres into the Bay of Bengal each year (Babel & Wahid 2008). Himalayan glaciers, primarily Gangotri and Satopnath, provide an estimated 30–40 per cent of the water in the Ganges, and the rest comes from melting snow and monsoon rains (Sharma & Sharma 2008). Glaciers in the Kailash Range of the northern Himalayas are the main sources of water for the Brahmaputra River in the dry season. After it moves out of Himalayan range, the Brahmaputra gets heavy rainfall, ranging from 2,500 millimetres to 6,400 millimetres per year (Singh & Arora 2007).

The Ganges River is the lifeline for the two most populous provinces of India: Uttar Pradesh and Bihar. The water of this river is used for irrigation, drinking water, and industrial purposes. The river is also used as an important navigation route. The Brahmaputra River – the biggest trans-Himalayan river system with the highest average annual runoff – is largely unexploited in Indian side as it flows through the hill areas of its northeastern part. The Brahmaputra, in the monsoon season, regularly causes severe floods in the Indian province of Assam and Bangladesh. The rapid deforestation of the Himalayas has also enhanced severity of the floods in the basin. The Ganges and Brahmaputra provide irrigation, navigation, fishing habitats, and water for domestic and industrial use, and also their flows resist salinity intrusion into Bangladesh and protect the environment and ecology of the southern region of Bangladesh. Nearly 80 per cent of the Ganges' annual flow takes place from July to October. This monsoon flow is not only sufficient, but it also regularly creates floods in India and Bangladesh. It is predicted that climate change will have serious implications for flood discharges in Bangladesh (Mirza, Warrick, & Ericksen 2003). In recent years, frequent extreme floods on the Indian side are also being increasingly blamed on climate change (Mallet 2013). Though the basin is witnessing increasing extreme flooding in

monsoon seasons, the dry-seasons runoff is not adequate to meet the requirements of these regions. A scientific report commissioned by the World Bank
Group, analyzing existing global warming trends, warns that by the 2040s, India
will see a significant reduction in crop yields due to reduced water availability
(Schellnhuber et al. 2013).

Basin without a river basin institution

This most populated river basin lacks any basin or sub-basin-based initiative
to address challenges regarding water sharing issues, particularly when global
climate change threatens to reduce its runoff and increase the demand for water
extraction. At best, the riparian countries have managed only to agree bilaterally.
The most bitter and protracted river disagreement has been between Bangladesh and India over the sharing of the flow of the Ganges. The disagreement
came up between India and Pakistan in 1951, when India expressed its desire to
build a barrage at Farakka, 18 kilometres upstream from the East Pakistan (later
Bangladesh) border. The proposed project included a 38-kilometre link canal of
40,000 cusecs capacity, to take off from the barrage to augment the waters of the
Bhagirathi-Hooghly at the lower point. The official reason given for this water
diversion scheme was to flush out the silt and to keep the Calcutta port navi-
gable. In spite of Pakistani objection, India took unilateral decision to start the
barrage construction in 1962. The Farakka Barrage became operational in 1975,
and in January 1976 India began diverting water at Farakka, which prompted
Bangladesh to raise the issue in various international forums.

In the dry seasons, the average minimum runoff at Farakka was estimated in
1975 at only 55,000 cusecs. From this, India wanted to divert 40,000 cusecs with
the help of a diversion canal at Farakka, while Bangladesh demanded all 55,000
cusecs for its own uses. The increasing upstream withdrawal for the irrigation
purposes in Uttar Pradesh and Bihar had further reduced the dry-season flow
at Farakka. From 1994, Bangladesh complained of getting only 9,000 cusecs in
the most acute dry seasons, which leads to the assumption that the water availability at Farakka has come down to at most 49,000 cusecs in the dry seasons.
This new figure created a further hurdle for the negotiators to reach an agreement (Swain 2004). However, in December 1996, the Prime Ministers of India
and Bangladesh signed the Ganges River water sharing agreement. Instead of
usual short-term agreements to share the dry-season flow at Farakka, both the
countries opted this time for a 30-year arrangement.

The Treaty stipulates that below a certain flow rate, India and Bangladesh each
will share half of the water. Above a certain limit, Bangladesh is guaranteed a
minimum level and if the water flow increases to a given limit, India will receive
a stipulated amount and the balance will be given to Bangladesh, which will be
more than the 50 per cent of the flow. The 1996 agreement was based on the
flow average of 1949–1988, but the real flow at Farakka in the 1990s was much
less than that. To get a reliable figure, the water experts should have taken the
average of the flow of last 10 years to the agreement. Unfortunately, the very

first year of the treaty witnessed a severely low dry-season run-off in the Ganges River.

The water sharing agreement came under serious threat because of less water flow in 1997 in the Ganges compared to the previous years. With the help of the political support, the 1996 Agreement withstood the challenge. However, the dry season runoff of the Ganges River has improved since then, due to good rainfall in the upper basin areas, and possibly increasing the melting of snow in the Himalayan glaciers as result of climate change. But, if the upper basin areas face a drought cycle, it will be very difficult for the agreement to satisfy both parties. The reduced flow in the Ganges system has potentially wide-ranging socioeconomic and environmental implications for Bangladesh (Mirza 1997).

The most important contribution of the 1996 Ganges Water Sharing Treaty is that it provided an encouraging bilateral environment for discussing and deliberating on a number of other river water-sharing issues. This Treaty itself requested addressing issues like flood management, irrigation, river basin development, and hydropower generation for the mutual benefit of the two countries. Bangladesh has recently started taking actions against Indian separatist rebels who have taken refuge there. Both countries have signed new extradition and visa agreements and have established stronger economic ties. However, there is unfinished business between the two countries. India has not yet signed the Teesta Water Treaty, which will provide Bangladesh its share of water from the Teesta River, which flows from India. Meanwhile, India is asking for Bangladesh to allow transit access to its landlocked northeastern states, which are separated by Bangladesh.

The challenge before the two most populous countries in the sub-continent – India and Bangladesh – is to provide their growing populations with access to clean drinking water, reliable irrigation, cheaper energy resources, and flood protection. There is a sign of growing bilateral cooperation in recent years to share and develop common river water resources. This cooperation has been achieved in spite of strongly critical historical and ethnic political factors. Increasing water scarcity and its direct, adverse effect on the agricultural sector – coupled with political opportunities – has paved the way for cooperative water management of shared rivers. However, global climate change (which raises the possibility of severe water flow fluctuation in the river systems), changing patterns of water demand, and a growing need in the basin raise serious apprehension for successful water cooperation in future. In spite of recent controversies involving climate science, there is no uncertainty about the occurrence of climate change. But at the same time, there is a great deal of uncertainty about the future projections over the intensity of climate change and predictions about its impacts. This confusion further complicates management of shared rivers between India and Bangladesh.

India and Bangladesh are hydrologically inter-linked for having 54 rivers in common. And it's not only the Ganges; several other common rivers have been the source of conflicts and tension between two neighbours. Being a lower and weaker riparian, Bangladesh is keen on having water-sharing agreements on all the common rivers. After the Ganges Water Treaty of 1996, Bangladesh has attached highest priority to the sharing of the Teesta River. The other rivers in discussion for sharing are the Manu, the Khowai, the Gumti, the Muhuri, the Jaldhaka (Dharia), the Torsa (Dudhkumar), and the Feni.

Both countries have an ad hoc water allocation over the Teesta River since 1983. Water-sharing has again become a source of dispute as Bangladesh demands a guaranteed flow in the dry season due to its increasing water demand (Wirsing & Jasparro 2007). Both countries are in negotiation over the increasingly scarce water of the Teesta River. India's plan to build a Tipaimukh dam is another bilateral issue of increasing concern. The project has been on the drawing board for a long time, but in July 2009, National Hydro-electric Power Corporation (NHPC) was entrusted to implement it. The dam is to be constructed 500 metres downstream from the confluence of the Barak and Tuivai Rivers in Manipur. The main objective of the Tipaimukh dam project is to generate 1,500 MW hydropower and flood control on 2,039 square kilometres of area. India argues that the dam is only meant for hydropower generation and flood control, so it will not affect the quantity of water flow reaching downstream in the dry season. Bangladesh, however, is not satisfied with this explanation and demands to be consulted in the planning of this project. Opponents of the Tipaimukh dam argue that the project will virtually dry up Bangladesh's Surma and Kushiara Rivers, and will disrupt agriculture, irrigation, fisheries, drinking water supply, navigation, and ground water levels. Besides political opposition in Bangladesh, civil society and environmental groups in both India and Bangladesh are critical of this project due to the possible adverse ecological impact of the dam. However, governments of both countries expect the construction of the dam to solve the flood problem caused by the Barak River in Assam and in parts of Bangladesh.

The other Indian project – the proposed massive river linking plan – keeps Bangladesh tense, as it fears that the implementation of the project will inflict serious damage to its economy and environment. Since 1982, India has carried out detailed studies for inter-linking its major rivers. In December 2002, the Indian government constituted a task force to carry out the feasibility studies for the linking of rivers by the end of 2016. The idea is to transfer water from river basins with 'excess' water to deficient river basins via 260 links (Alagh, Pangare, & Gujja 2006). India also contemplates transferring some of the Brahmaputra water to the peninsular rivers, like Mahanadi in Orissa and Godavari and Krishna in Andhra Pradesh. Indian strategy on working bilaterally with Nepal for the augmentation of the Ganges water is perceived by Bangladesh as India's desire to utilise the increased water supply of the river to meet its own growing demand. Besides adverse environmental effects, these inter–basin water transfer schemes will be politically unfathomable within India. Any attempt to carry out the plan from surplus to deficit basins will not only infuriate the existing river sharing disputes, it will also likely introduce many new conflicts between various provinces of India (Swain 2010a).

Bangladesh is always apprehensive of any Indian plan for developing the water resources of the shared rivers. Since 1978, Bangladesh has been reluctant to accept any plan to augment the dry-season flow of Ganges with the 'surplus' waters from the Brahmaputra River, particularly after India's growing interest in its grand river linking plan. In 1978, Indian suggested connecting Brahmaputra with the Ganges through the construction of a 320 kilometre link-canal, 120 kilometres of which would run through Bangladesh. This 75 metres wide and 10 metres deep canal would have a carrying capacity of 100,000 cusecs and would be supplemented by three storage reservoirs on the Brahmaputra River. The plan was justified on the grounds that the waters of the Ganges were inadequate to meet the increasing demand, whereas Brahmaputra, which has four times more dry-season flow than Ganges, runs down to the Bay of Bengal unused.

Many reasons prevent Bangladesh from accepting this Indian plan. Bangladesh is apprehensive of its water being diverted from the Brahmaputra River, which will cause water scarcity and ecological problems. The proposed link-canal would also add to Bangladesh's already existing communication problems by dividing the northern part of Bangladesh from the rest of the country. It would be a very high-cost project, and would displace millions of people in a highly populated country. Bangladesh rejected the Indian Plan in 1978 and argued for a tripartite partnership of herself, India, and Nepal in seeking a solution to the water scarcity problem in the Ganges basin. The Bangladeshi proposal envisages seven storage dams on Ganges tributaries in Nepal. These dams would store monsoon flow and yield a year-round flow of about 176,400 cusecs of water. The Bangladeshi view of the best means of augmenting the Ganges dry-season flow lies within the Ganges system itself.

The Bangladesh position is that the storage dams on the Nepalese tributaries would also help to control the flooding in the Ganges during the monsoon seasons. However, Bangladesh desires the involvement of Nepal in Ganges River management, since it would be more difficult for India to break a trilateral commitment than a bilateral one. Bangladesh also has a couple of tempting offers for Nepal to be included, like exporting large amounts of hydro-power via an alternative navigation route to land-locked Nepal by digging a 30 kilometres-long canal across Indian territory to link the rivers of both countries. This would give Nepal a direct outlet to the sea.

India was quite averse to addressing the water-sharing issue at the basin level. After years of waiting and the hope of getting Brahmaputra water to its mainland, in 1986 India deflected from her earlier stand by withholding her objection to the inclusion of Nepal in the water-sharing arrangement (Gulati 1988). Though Nepal had expressed its intention to participate in the arrangement since 1977, it made no serious effort to prepare a cost-benefit analysis (Verghese & Iyer 1993). The realisation of massive cost came only after India's acceptance of her inclusion, and then Nepal started to vacillate on the plea of studying the proposal. Past debates over bilateral water development projects with India forced Nepal to think twice about the Bangladeshi proposal at that time. The prospect of a navigation route has not been able to cut much ice because Nepal would remain at the mercy of India anyway, as the proposed canal would be in Indian territory. After the failure of joint approach to Nepal, the matter was not followed up seriously.

While negotiating the December 1996 treaty with Bangladesh, India stressed the need of permanently solving the Farakka issue while completely ruling out the involvement of Nepal. India favours a long-term settlement of Ganges water sharing and maintains pressure on Bangladesh to accept its Brahmaputra link-canal proposal. Bangladesh, on the other hand, still hopes for Nepal's participation in the water management of the Ganges-Brahmaputra basin. Before the Indian prime minister's visit to Bangladesh in September 2011, the Bangladeshi Prime Minister's Foreign Affairs Advisor Gawher Rizvi, in an interview to Press Trust of India, mused, 'We hope during the meeting of the two Prime Ministers the possibility of the joint river management will be extensively looked into . . . Which in the longer run should engage other co-riparian nations' (OutlookIndia. com 2011). Bangladesh continues to voice for joint collective management of shared river resources and hopes for Nepal to participate in this collaborative venture. However, Nepal has reasons to be cautious.

India and Nepal, though, have relatively good bilateral relations; both have had many rounds of occasional tense negotiations relating to hydropower generation, irrigation water, and flood control. Several early agreements about shared water projects have been very controversial in Nepal. The water sharing of the major rivers originating in Nepal and flowing into India has strained the relationship between the two countries. Nepalese feel that they have not been treated equitably under the various water-resource development agreements with India, including Sarada, Kosi, and Gandak (Upreti 1993). Negotiations regarding projects on the shared river systems have been dominated by controversies due to a lack of mutual trust (Salman & Uprety 1999). A serious dispute over the river water issue came up in the early 1990s. In December 1991, both countries signed a Memorandum of Understanding to construct a barrage at Tanakpur, for which Nepal agreed to provide 2.9 hectares of land. Nepalese rivers have tremendous potential for hydropower generation. However, this issue became extremely controversial in Nepal due to the internal political situation, which created bilateral tension. After several rounds of negotiations, the Prime Ministers of India and Nepal signed the Mahakali Treaty in February 1996. Though this treaty on the integrated development of the Mahakali River came up as a solution to the legacy of disagreement between Nepal and India over the Tanakpur Barrage project, it has not silenced the opponents in Nepal (Gyawali & Dixit 1999). Nepal lacks the capital and technology required for such large projects, and also needs a buyer for the surplus hydropower. Due to geo-strategic reasons, India is probably the only country that could provide economic and technical assistance. Thus, India's direct involvement in the utilisation of the river water in Nepal is crucial, and that contributes to regular controversies. This also inhibits Nepal to be a willing partner in the joint management of Ganges-Brahmaputra water resources.

Despite huge water resources available in the region, the inability among and between countries of the basin to reach mutually beneficial comprehensive agreements could invite more conflicts in the days ahead, particularly in the context of climate change. The way in which rivers are used in one country can indeed have far-reaching effects on nations downstream. India's proposal to link major rivers in the region to provide water to its arid provinces is causing anxiety among its neighbours, particularly Bangladesh and Nepal. Beside the Indus River Treaty, in which World Bank played a decisive role, the South Asian states have generally adhered to the strict practice of bilateral negotiation. This bilateralism enhances the importance of past disagreements, and suffers from a perception that limits the needs of neighbouring states. Moreover, countries in this region have always preferred the large water development scheme within the state-centric approach. Moreover, river water-sharing issues have, in most cases, been negotiated in isolation, failing to link other associated issues for a better and successfully negotiated outcome. However, this business-as-usual approach is being increasingly challenged by a number of new developments. One of them is the threat of looming climate change, which is threatening even the survival of this sacred river (Wax 2007). The climate change predictions raise serious risk of increased fluctuation in the dry-season water flow of the river, and that might pose dangers to the ongoing water sharing arrangement between Bangladesh and India in the long run (Srabani 2010).

GCC and the basin

The quantity and nature of runoff is projected to change substantially in the Himalayan rivers as a result of climate change. While uncertainty remains regarding the accuracy of various climate change predictions, forecasts indicate that changes in climate will further exacerbate the existing variability of water flow in the rivers. The southwest monsoon system in South Asia is the most important climatic phenomenon. Shifts from the normal regime can have catastrophic effects in the form of floods, droughts, or famines in the region. Though it has been predicted that there will be an impact on the rainfall pattern of the country due to climate change, it is not yet certain as to what that will be exactly, with some models predicting increase and others showing a decreased rainfall pattern. Changes in the spatial and temporal distribution of precipitation and temperature are expected to interact in complex ways that may change the equilibrium and attributes of runoff that reach rivers. In the Ganges-Brahmaputra basin, climate change is expected to increase temperatures, resulting in the retreat of glaciers; increased alteration in precipitation pattern, ensuing in greater degree and rate of recurrence of droughts and floods; and even lead to rising sea-levels (Cruz et al. 2007). It is still early to properly grasp the projected impacts of climate change on groundwater recharge within the Ganges-Brahmaputra basin; however, declining rates of recharge would affect groundwater irrigation systems throughout the region. Seasonal and lasting drying up of wetlands, lakes, and streams – and reduced seasonal stream flow – can have considerable effects on water availability. The basin is likely to experience severe effects from climate change due to the large population and the high dependence on irrigated agriculture (Immerzel, Van Beek, & Bierkens 2010).

The increasing uncertainty over the availability of water from rainfall, river runoff, and groundwater recharge due to impacts of climatic change poses a serious threat to the food security in the region. Because of high dependence on the monsoons – as they provide more than two-thirds of annual precipitation in a period of four months – the Ganges' basin climate has the highest seasonal concentration and variability of rainfall in the world. If some of the projections about monsoonal precipitation are indicative of future trends, the associated risks are extremely high. Drought is a chronic problem in Bangladesh. It is mostly caused when the monsoon rains, which normally provide most of Bangladesh's annual precipitation, are significantly decreased. Some climate models project decreased annual precipitation for Bangladesh, and the projections even tend to demonstrate particularly reduced rainfall in the winter months. Changes in rainfall patterns, natural water storage, and surface water runoff from the Himalayas pose challenges for the operation of ongoing surface water irrigation systems. In this case, there is going to be increased possibility of more droughts in Bangladesh in the future (Agrawala et al. 2003). Besides adversely affecting agricultural production, climate change may also reduce the inland fisheries sector, which provides livelihood to a large fisher community (Das et al. 2013).

However, not so unexpectedly, the estimates from the climate models do not provide a clear picture about the future drought situation. It is a fact that the calculated changes in precipitation in this region are not very significant. The overall predictions of global climate change, especially the emphasis on shifts in mean temperature and its impact on glacier thinning, do not take into account regional complexities in the mountains related to the effects of topography and elevation. If that is taken into account, as some studies suggest (primarily due to precipitation changes and further glacial melt), increased runoff is expected in the major river basins shared between India and Bangladesh (Milly et al. 2005). However, warming climate may increase the availability of surface runoff of these rivers through the melting of glaciers in the short and medium term till 2050, but in the long run the supplies will be reduced (Barnett et al. 2005). The size of the glaciers in the Himalayas has reduced from 2,077 square kilometres in 1962 to 1,628 square kilometres in 2007, an overall reduction of 21 per cent (Kulkarni et al. 2007). Rapid glacial retreat is associated with the highest levels of black-carbon concentrations around the margin of the Tibetan plateau. This creates a layer of warm air to become a major catalyst of glacier melt (Morton 2011).

Although data are limited and contested, there is a general agreement that the Himalayas are particularly vulnerable to the effects of global climate change. Rising temperatures will precipitate glacier and snow melt, modifying river regimes and increasing risks of floods. Xu et al. argue that 'on-going climate change over succeeding decades will likely have additional negative impacts across these mountains, including significant cascading effects on river flows, groundwater recharge, natural hazards, and biodiversity; ecosystem composition, structure, and function; and human livelihoods' (Xu et al. 2009). This increased rate in glacier melting may add to the runoff of the rivers for some time, but at the same it will also bring more snow avalanches and glacial lake outburst floods (GLOFs) at high elevation. This phenomenon will not only pose a risk for the existing dams and other projects at the upper reaches of the river systems, but will also make it difficult to plan any hydro projects in future. Glaciers and snow-covered areas of the Himalayan Mountains are an important source of freshwater storage. The impacts of climate change pose serious risks for the future freshwater storage at high elevations and freshwater runoff to the rivers downstream.

There is enough water in the shared rivers in the wet season for the both Bangladesh and India to argue over the supply, and flood control becomes their priority. But, during the dry season months (December through May), water supply dwindles in these rivers and ice- and snowmelt from the Himalayan Mountains is critical. Water supply from melting ice and snow also help the basin to cope with water scarcity in drought years. The Ganges River alone receives approximately 9 per cent of its flow from the melting snows of the Himalayas (Jianchu et al. 2007). However, climate change accelerates glacial retreat and shifts the pattern of ice- and snowmelt. This poses serious challenges for India, Bangladesh, and Nepal to manage their shared river water, both in the short and long run. It has been argued that climate change will have a far greater and more damaging effect on the Ganges River than other human-made factors, such as dams and diverting water for irrigation (*The Guardian*, 22 April 2009). In the short run, increased snow melting threatens existing water storage projects and increases the possibility of flash floods, landslides, and avalanches; the long-term scenario is extremely bleak as it may transform these mighty perennial rivers to seasonal ones, which will have huge implications in economic, political, and ecological fronts.

Riparian response to GCC: adapt, resist, or subvert?

Perceived short-term national interests of the Ganges-Brahmaputra basin countries often clash with the environmental and ecological needs of the river system. Smaller riparian states, Nepal and Bangladesh, do see the necessity of establishing basin-based institutions to address the emerging threats of climate change to their water resources. However, there is certainly a lack of political will in the basin to set up effective, basin-based institutions. The trust deficit is substantially high among the basin countries. Water sharing and management issues do not respect national boundaries, and in many cases they require basin-based solutions. The threat of climate change has become a major concern, at least officially for the basin countries in recent years. However, basin countries suffer from a serious climate of non-cooperation, and they continue to show this lack of collective concern and seriousness towards threats posed by climate change (Swain 2011). Though the basin countries have not come together to take a joint approach, they are certainly aware and regularly admit their vulnerability to climate change-induced water challenges.

Water availability in the Ganges-Brahmaputra basin is driven by monsoons, melting snow, and glaciers from the Himalayas. Flooding takes place regularly in this basin and causes enormous destructions to lives, property, crops, and infrastructure. Future climate scenarios from the global climate models project that the frequency, magnitude, and extent of flooding in the region may increase due to climate change. Increased flooding will have serious implications for food security and impact public health. Crops in the basin are vulnerable to climate variability and extremes. Crop damage results mostly due to heavy rain-induced flooding, flash floods, and riverine flooding. Crops are also damaged by droughts, cyclones, and associated surges and hailstorms. Climate change can also affect public health in many ways. Diseases such as diarrhoea, dysentery, and dengue are water related and prevail in the flood season. Moreover, human displacement as a result of floods has an impact on public health as well (Mirza 2011).

With the rapidly growing population in the basin, the social vulnerability has grown, and it forces people to farm in fertile areas around the rivers that are prone to regular flooding. These people and their assets are highly vulnerable to the increasing number of floods. However, regular flood forecasting in the basin, particularly in Bangladesh, has only a short time horizon of one to two days, using simple statistical methods to forecast a flood. This is, however, not enough time to prepare for the flood. Due to lack of data from upstream countries, flood forecasts are difficult to calculate. India, for instance, provides very minimum upstream data to Bangladesh. India, in fact, barely has a centralised, nation-wide water data system for its own regions. Data is either unavailable because it is not properly measured, or is not made available, owing to transboundary political reasons. As Webster et al. estimate, a seven-day flood forecast in Bangladesh has the potential of reducing the post-flood household and agriculture costs by 20 per cent, compared to 3 per cent for a two-day forecast (Webster et al. 2010).

Besides severe flooding in rainy seasons, the Ganges-Brahmaputra River is also already under a great deal of stress during dry seasons, due to excessive water withdrawal or unsustainable agricultural development. The overlapping effect of climate change and the impact of dams, barrages, and other human infrastructure may be exacerbated by this stress. Some also predict political tension as result of climate-induced migration in the basin. Bangladesh, a highly populous country, is in particular threatened by devastating floods or other damage by monsoons, melting glaciers, and tropical cyclones. Climate migrants from Bangladesh will create political tension as they cross the border, particularly to India. Not only in Bangladesh, devastating effects of climate change in poor Nepal will displace a large number of its population and will put further stress on the fragile peace of the country (Podesta & Ogden 2008). Therefore, anticipating and planning adaptive strategies to address climate change in the basin may be critical (Palmer et al. 2008). Taking into account their adaptive capacity, Babel and Wahid evaluate the vulnerability of the Ganges-Brahmaputra basin countries: 'Overall vulnerability of water resources is the highest in Bangladesh, (. . .) followed by Nepal and India. Bangladesh is also the most sensitive country among the three, while Nepal has the least capacity to adapt' (Babel & Wahid 2011: 354). However, the sources for the vulnerability of these countries are not the same. In India, the stress comes from higher and increasing water demand and withdrawal. In Bangladesh, it is high seasonal fluctuation in water availability and rapidly declining water quality. Nepal suffers from delicate geo-ecological sensitivity. However, all the basin countries in varying degree face governance challenges; this further enhances their water vulnerability.

Ganges-Brahmaputra basin countries are making some attempts to allocate their limited and scarce resources towards adapting to challenges of climate change. These preparations are a combination of 'hard' and 'soft' measures to protect their population and assets. 'Hard' measures include water infrastructure, emergency shelters, early warning systems, and disaster relief operations. On the other hand, 'soft' measures include design standards and building codes for infrastructure and countercyclical shifts in cropping patterns to offset agricultural losses (Dasgupta et al. 2010).

Bangladesh

Poverty and geography has placed Bangladesh on 130th position out of 139 countries for its network of roads, power, and ports, according to the *2010 World Economic Forum's Global Competitiveness Report*. Climate change poses significant risks for the infrastructurally weak Bangladesh, particularly to its water and coastal resources in terms of certainty, urgency, and severity of impact (*The New York Times*, 28 March 2014). This situation has made Bangladesh, as *The Huffington Post* (23 October 2013) puts it, as the 'poster child' of climate change. Besides rising sea levels and increasing numbers of cyclones, Bangladesh is extremely vulnerable to increasingly unpredictable monsoon and glacial melt. The usual approach to river water management in Bangladesh has been based on structural interventions in flood control, drainage, and irrigation. Thanks to international support, there is a developing trend towards creating a flood warning system, flood proofing, and adopting responses to hazardous conditions. Earlier efforts were mainly concentrated on the optimal utilisation of water resources, but environmental consideration and efficient water management infrastructures were not in fashion. There is a changing attitude, particularly after the adoption of National Water Policy (NWPo) and National Water Management Plan (NWMP), Bangladesh is striving towards sustainable water resources management through incorporating eco-efficient infrastructural and non-structural measures (Ahmed 2011).

Due to climate-related risks on monsoon flooding over time, the government of Bangladesh has highlighted severe monsoon floods as a significant hazard and ensuring adequate flood protection infrastructure as a 'pillar' of its Climate Change Strategy and Action Plan of 2009. In the last two decades, through several government-donor partnerships, Bangladesh has undertaken some infrastructural development to adapt to climate change, e.g., dredging of waterways and flow regulations on coastal embankments. In recent years, training and creating awareness among farmers on climate change is one of the key components of adaptive strategy being adopted (Program Support Unit 2009).

There is no doubt that the agricultural sector in Bangladesh is highly vulnerable to climate variability and extremes. Crop damage usually occurs due to heavy rainfall-induced flooding, flash foods, riverine flooding, and moisture stress during a drought. Crops are also damaged by cyclones and associated surges and hailstorm. Climate variability, change, and extreme weather can certainly create food insecurity for the poor in Bangladesh. In recent years, the government of Bangladesh has given its policy-planning attention to coastal planning and land-use management. It has adopted Land Use Policy (2001), Coastal Zone Policy (2005), Tsunami Vulnerability Map (2005), and Coastal Development Strategy (2006) (Ahmed 2011). However, due to economic and governance challenges, there is a huge gap in implementation of policy and available strategy directives. Moreover, the review of existing policy documents of Bangladesh suggests that, besides agricultural sector, the policies in general have not paid sufficient attention to other sectors, like fish and livestock production and forestry, which might have potential impacts from climate change hazards.

Nepal

Since the mid-1990s, Nepal is suffering from acute political instability. Nepalese government's main priority until 2006 was to control the Maoist insurgency group. On 21 November 2006, a Comprehensive Peace Agreement was signed by the government of Nepal and the Communist Party of Nepal (Maoist), but it did not bring anticipated peace immediately. Popular opposition resulted in the overthrow of the king in 2008 and Nepal managed to hold the election that year, but political stability in the country is still far from being achieved.

Political instability and extreme poverty pushed Nepal to passively resist bringing adaptive policy changes to the climate-change induced challenges to its rich water resources. Only in early 2010 did the government of Nepal launch the Climate Change Management Division. After few months, the government endorsed the National Adaptation Program of Action. Moreover, a multi-stakeholder Climate Change Initiatives Coordination Committee (MCCICC) was also formed to implement collaborative programs coordinated among donor governments, international institutions, and international and national non-governmental organisations (Pandey 2011). With donor support, Nepal has taken steps towards establishment of the Climate Change Knowledge Management Centre and is preparing a climate change strategy to operationalise the Climate Change Policy, which focuses on climate adaptation and resilience.

Nepal has huge potential for hydropower development and the present production is approximately 600 MW, which is less than 2 per cent of its economically feasible potential of 42,000 MW. In 2012, Nepal has announced an ambitious economic development plan in which focus has been on new hydropower projects. However, this plan takes climate change challenges into superficial consideration. In 2011, with USAID support, Nepal has introduced a five-year program, *Hariyo Ban*, to mitigate adverse impacts of climate change and threats to its biodiversity. This program aims to build lasting resilience against climate impacts, and reduce deforestation and land degradation. World Bank also provided a loan to Nepal for the implementation of Pilot Project for Climate Resilience (Bhusal 2011). However, in November 2011, the Asian Development Bank, in its report *Nepal: Mainstreaming Climate Change Risk Management Development,* concluded that Nepal is at a very early stage in addressing the impacts and risks of climate change, merely at that of bringing local and national awareness (Asian Development Bank 2011). Planning and policy making in key sectors like water and agriculture are slowly taking risks associated with climate change into consideration. Nepal suffers from insufficient financial resources and serious governance challenges to effectively integrate climate change risk management into its water development planning.

India

As the world's third largest economy (in purchasing power parity terms), and the fourth largest emitter of greenhouse gases, India, like many other large developing countries, feels strongly that it is not responsible for the threat of climate change that has been created – they believe, instead, that unsustainable consumption patterns of rich, industrialised countries are responsible for it. India is already experiencing impacts of climate change, including water stress, drought, storms, and flooding, and related negative consequences on health and livelihoods (National Intelligence Council 2009). While it recognises that global warming will affect the country seriously, the government is of the view that the process of adaptation to climate change must have priority, and that the most important adaptation measure is development itself. The government claims it is already spending over 2 per cent of gross domestic product (GDP) on measures to adapt to the impacts of the changing climate, and it estimates that climate change could result in a loss of 9–13 per cent of the country's GDP in real terms by 2100 (Mehra 2009). India takes a formal political position against developed countries on the climate change issue, but officially accepts the threat of climate change to its own development and peace, and makes some attempts to adapt to the new challenges.

In June 2008, India's Prime minister released the National Action Plan on Climate Change (NAPCC). The NAPCC outlines a strategy by which India will adapt to climate change, while maintaining a high growth rate, protecting the poor and vulnerable sections of society, and achieving national growth objectives. It focuses on eight areas intended to deliver maximal benefits to development and climate change (mitigation and adaptation). However, detailed action plans for each mission, and any clear targets and timetables, are still not available. India can claim some credit for a few achievements that provide a good foundation for further improvements to limit its emission of greenhouse gases in future. For example, India was the first country to establish a ministry for non-conventional energy sources and has the world's fourth largest installed wind power capacity.

India has mostly focused on the climate change issue in the energy sector, while the water resource sector has not received the adequate attention as it deserves. The National Water Mission, a part of the National Action Plan on Climate Change, identifies various threats to water resources in India due to climate change. It predicts expected decline in the glaciers and snow fields in the Himalayas; increased drought-like situations due to the overall decrease in the number of rainy days in large part of the country; increased flood events due to the overall increase in the rainy day intensity; effected groundwater quality in alluvial aquifers due to increased flood and drought events; impacted groundwater recharge due to changes in precipitation and evapotranspiration; and increased saline intrusion of coastal and island aquifers due to rising sea levels (National Water Mission 2009). In spite of acceptance of serious challenges, the Government of India has only initiated some policies to improve investment and management of water in the context of climate change, which include: Accelerated Irrigation Benefits Program, Hydrology Project, setting up of Water Quality Assessment Authority, Command Area Development and water management program, National Project for Repair, Renovation and Restoration of Water Bodies directly linked to Agriculture, Flood Management, and River Basin Organizations. There is also an effort to provide comprehensive training and capacity-building schemes to the water officials at the various levels of the government to be climate-wise knowledgeable (Asian Development Bank n.d.). In 2009, India launched the new National Ganga (Ganges) River Basin Authority (NGRBA) to clean and conserve the Ganges River in India. The World Bank is also providing major support to this program. In 2013 only, India's Ministry of Water Resources took up studies of measuring basin-level climate change impacts of its major rivers, including the Ganges-Brahmaputra, applying Regional Climate Model (*Times of India* 18 June 2013). India has shown its interest by taking up some policies to maintain good water quality; however, the successful implementation of these policy responses to the climate challenge needs both a strong scientific knowledge base and efficient institutional support at the national, regional, and local levels. India's long-standing aim at diverting water from other basin countries to augment its own supply motivates her desire to resist undertaking any concrete measures to address climate change-induced water scarcity challenges.

Basin's response to GCC: far from adapting to the new challenges

For years, developmental and demographic changes have created new challenges to water security in the Ganges-Brahmaputra basin. Rapid economic and population growth, industrialisation, and urbanisation have posed daunting tasks for efficiently managing this important Himalayan River system. Climate change brings the possibility of increasing temperatures, more extreme weather events, droughts, floods, and changing precipitation patterns in the near future. All of these factors will have a serious impact on the water availability and quality in the basin. Without basin-based cooperation, the impact of these changes will have devastating effects on the lives of millions of people in the region – in particular, the downstream areas.

Despite the growing concerns of catastrophic implications of climate change on water demand and supply, the Ganges-Brahmaputra basin lacks meaningful institutions to deal with the crisis. In December 1996, India and Bangladesh signed the Ganges River water sharing agreement. Though the agreement wishes for an integrated management of the watercourses, the weakness of the treaty is that it does not include a clear guideline. This treaty refers to other water-related issues, like flood management, irrigation, river basin development, and hydropower generation for the mutual benefit of the two countries. The treaty also stipulates that below a certain flow rate, India and Bangladesh each will share half of the water. However, Bangladesh is particularly wary, as the reduced flow in the Ganges system has potentially wide-ranging socioeconomic and environmental implications for them (Mirza 1997). The long-term effect of faster glacier melting is going to be critical for the dry-season flow of the river, and that will pose serious problems for the bilateral agreement to reallocate the decreasing water in the face of increasing demand (Swain 2010a). Primarily due to India's reluctance, Nepal and Bhutan prefer not to show enthusiasm for multilateral cooperation, thus reducing the possibility for a basin-based management of the Ganges-Brahmaputra River in the future (Swain 2012).

India and Nepal, though, have relatively good bilateral relations; both have had many rounds of occasional tense negotiations relating to hydroelectricity generation, irrigation water, and flood control. Several early agreements about shared hydro-projects have been controversial in Nepal. Nepal lacks the capital and technology required for such large projects and also needs a buyer for the surplus hydropower. Due to geo-strategic factors, India is the only country that could provide assistance. Thus, India's direct involvement in the management of the river water in Nepal is crucial, and that contributes to regular controversies. On the other hand, India's relations with Bhutan – a minor upstream contributor to the basin –are much smoother. This is reflected in several hydropower development agreements signed by the two countries. Even on recent collaborative hydro projects, Bhutan and India have agreed to share the benefits of development. However, for its own interest, India prefers to deal with Bhutan and Nepal bilaterally, not on a common platform.

In recent years, there have been some cosmetic attempts to bring a sub-regional cooperation among Bangladesh, Bhutan, Nepal, and India to meet the increasing threats of climate change and to plan for adaptation strategies in the Eastern Himalayas region. Four countries held a Ministerial Level meeting on 19 November 2011 at Thimphu, wherein a regional 'framework of cooperation' was agreed upon. This 'Framework of Cooperation' aspires to promote regional cooperation to build resilience to climate change in the southern watersheds of the eastern Himalayas. There was also an agreement on a mechanism to implement the 'framework of cooperation' (Government of Bhutan 2011). However, this sub-regional initiative has seen very little progress at the implementation level. This is primarily attributed again to India's lack of interest in multilateral policy making in South Asia.

As Nepal's involvement is critical as an upstream nation for the Ganges part of the basin, in the Brahmaputra side, upstream riparian China holds the key. In Asia, most important rivers have their source in Tibet, where China has the de facto control over the water availability that impacts the life of 85 per cent of Asia's population, or 47 per cent of the world's population. Northern China is regularly struck by droughts, and due to the great uncertainties of climatic change, some climate scenarios suggest that the lives of millions will be in danger as the temperatures keep rising. In 2001, the government of China initiated a mega-project to divert water from the south to the dry regions in the north of the county. The South–North Water Diversion Project aims to divert 44.8 billion cubic metres water annually from the Yangtze River to the north. The construction of three diversion routes, stretching south to north across the eastern, central, and western parts will connect China's four main rivers – Yangtze, Yellow River, Huaihe, and Haihe (Wong 2011). China's thirst for water is not only limited to its own internal rivers. There is a long standing plan in Beijing to divert 200 billion cubic metres of water from the upstream part of international rivers in southwestern China, like Lancang-Mekong, Nu-Salween, and Yarlung Tsangpo-Brahmaputra to northern parts of China. High economic cost and technological challenges have delayed this water diversion plan, but China has started building a large number of hydropower dams on these rivers. Until recently considered the last great undammed river in Tibet, the Yarlung Tsangpo-Brahmaputra River has been the new focus of China's efforts to increase their hydropower supply. China has already started constructing five major hydropower dams on the river (Watts 2010).

The initiation of these new projects in China generates anxiety for India, as it brings new uncertainties for the water management in the Ganges-Brahmaputra basin. China does not commit to any formal agreements or legally binding instruments regarding its shared rivers. China's general lack of interest in international legal provision for sharing its water is voiced in the vote against the 1997 UN Convention on the Law of the Non-navigational Uses of International Watercourses. China's advantageous geo-strategic and economic positions in the basin allow them to unilaterally build projects on the Yarlung Tsangpo River, without taking into account the needs of its downstream riparian countries. China has no economic or political incentive to agree to manage its waters within a basin-based framework. China regularly makes it clear that it is not interested either justifying its action or joining a formal regime to address others' water issues.

India opposes China's attitude, but also behaves in the similar manner with downstream Bangladesh. According to a report prepared by India's government's own think tank, Institute for Defence Studies and Analyses, Tibet's waters should be seen as 'global commons', not for China alone. This report blames China for being inconsistent in sharing flood data and not providing information about its water transfer plan. However, with regard to Bangladesh, the same report suggests, 'Bangladesh cannot change its lower riparian position and will have to accept cooperative arrangements based on water sharing and not on water rights. India as the upper riparian has the responsibility to ensure that the equitable principles are fairly adhered to without undermining its own requirements' (Institute for Defence Studies and Analyses 2010: 56–57).

Adaptation to changing circumstances in the Ganges-Brahmaputra basin needs a comprehensive policy for integrated water resource management and requires significant collaboration across the sectors (Hosterman et al. 2012). However, to be sufficiently prepared to meet challenges of global climate change, the basin has no common knowledge base or analytical framework to explore options and facilitate cooperative planning, nor is there an effective institutional mechanism for basin-wide riparian dialogue and cooperation (World Bank 2010). Basin mostly lacks the incentives, comprehensive policy, and institutional framework for the efficient, sustainable, and equitable allocation of shared water.

Conclusion

Considering the enormity of the climate change challenges in South Asia, there is at present no paucity of collaborative efforts to explore effective response strategies. The World Bank is presently supporting the Abu Dhabi Dialogue process, which was originally initiated by London's International Institute of Strategic Studies in September 2006. This Dialogue process brings together experts from the seven Greater Himalayan countries – Afghanistan, Bangladesh, Bhutan, China, India, Nepal, and Pakistan – on a periodic basis. Experts from the region also meet at various international conferences. This is the only multilateral framework involving all the riparian countries of the Ganges-Brahmaputra basin that is engaged in an informal and consultative process, and aims to 'build trust via information exchange and the co-management of development projects' (Morton 2011: 128). Unfortunately, the World Bank has failed in its hope to bring South Asian countries within a common water-sharing framework in line with the Nile Basin Initiative. However, the rapidly unfolding climatic crisis requires an inclusive dialogue mechanism, bringing together wider stakeholders in the basin. There is now a strong need for setting up an institutional, systematic, and collaborative river water management mechanism in the basin.

Basin-based institutions can help to reduce uncertainty surrounding Ganges-Brahmaputra riparian relations, which has been the bane of the region for a long time. If successful, these institutions can possibly shift the focus from disconnected and short-term interactions of the riparian countries into a continuous relationship that has scope for future routine gains. To date, however, all these expectations, like in Jordan basin, have not ventured beyond political statements and setting up lofty goals. However, basin-based water management institutions, if established, can play a very crucial role in the adaptation of climate change-related challenges to the region.

Because of to the impacts of climate change, droughts and floods are expected to become more severe and less predictable. To address these extreme challenges, there is an urgent need for basin-based cooperation, in order to augment the irrigation capacity in low-season periods and to decrease flood peaks in the monsoon months. However, there is certainly a lack of political will in the basin, particularly in India, to provide the space for multilateral river institutions to effectively emerge. As we witness in the Jordan basin, the trust deficit is extremely high in the Ganges-Brahmaputra basin, and many consider India to be the regional bully. India's refusal to take up any river management issue in a multilateral forum reduces the scope for possibility of any basin-based organisational framework to emerge. The advent of China in the basin's hydro-politics, like in the Mekong and the Nile, has brought further complications and uncertainties.

References

Agrawala, S., Ota, T., Ahmed, A. U., Smith, J., and Aalst, M., 2003, *Development and Climate Change in Bangladesh: Focus on Coastal Flooding and the Sundarbans.* Organization for Economic Co-operation and Development.

Ahmed, A., 2011, 'Some of the major environmental problems relating to land use changes in the coastal areas of Bangladesh: a review', *Journal of Geography and Regional Planning,* 4(1): 1–8.

Alagh, Y. K., Pangare, G., and Gujja, B. (eds.), 2006, *Interlinking of Rivers in India: Overview and Ken-Betwa Link.* New Delhi: Academic Foundations.

Alex Mifflin 23 October 2013 Bangladesh Is Drowning Because of Climate Change. The Huffington Post. Retrieved 1 March 2015 www.huffingtonpost.ca/alex-mifflin/bangladesh-climate-change_b_4150220.html

Asian Development Bank Technical Assistance Report, 2011, *Nepal, Mainstreaming Climate Change Risk Management and Development.* CDTA Project Number 44168, November 2011.

Asian Development Bank, n.d., *Water Resources Development in India: Critical Issues and Strategic Options.* Manila: ADB.

Babel, M. S., and Wahid, , S. M., 2008, *Freshwater Under Threat: South Asia.* Asian Institute of Technology: United Nations Environmental Program.

Babel, M. S., and Wahid, S. M., 2011, 'Hydrology, management and rising water vulnerability in the Ganges–Brahmaputra–Meghna River basin', *Water International,* 36(3): 340–356.

Barnett, T. P., Adam, J. C., and Lettenmaier, D. P., 2005, 'Potential impacts of warming climate on water availability in snow dominated regions', *Nature,* 438: 303–309.

Bhusal, R. P., 2011, 'Loan to fight climate change', *The Himalayan,* 24 January 2011.

Cruz, R. V., Harasawa, H., Lal, M., Wu, S., Anokhin, Y., Punsalmaa, B., . . . Hanson, C. E. (eds.), *Climate Change 2007: Impacts, Adaptation, and Vulnerability; Contribution to Working Group II to the Forth Assessment Report of the Intergovernmental Panel on Climate Change.* Cambridge, UK: Cambridge University Press: 469–506.

Das, M. K., Sharma, A. P., Sahu, S. K., Srivastavaa, P. K., and Rej, A., 2013, 'Impacts and vulnerability of inland fisheries to climate change in the Ganga River system in India', *Aquatic Ecosystem Health & Management,* 16(4): 415–424.

Dasgupta, S., Huq, M., Khan, Z. H., Masud, Md. S., Ahmed, M.M.Z., Mukherjee, N., and Pandey, K., 2010, *Climate Proofing Infrastructure in Bangladesh The Incremental Cost of Limiting Future Inland Monsoon Flood Damage.* The World Bank Development Research Group Environment and Energy Team, November 2010.

Gardiner Harris 28 March 2014 Borrowed Time on Disappearing Land. The New York Times. Retrieved 1 March 2015 www.nytimes.com/2014/03/29/world/asia/facing-rising-seas-bangladesh-confronts-the-consequences-of-climate-change.html?_r=0

Government of Bhutan, 'Ministry of Agriculture and Forest', 2011, *The Bhutan Climate Summit for Living Himalayas,* 21 November 2011.

Gulati, C. J., 1988, *Bangladesh: Liberation to Fundamentalism: A Study of Volatile Indo-Bangladesh Relations.* New Delhi: Commonwealth Publishers.

Gyawali, D., and Dixit, A., 1999, 'Mahakali impasse and Indo-Nepal water conflict', *Economic and Political Weekly*, 34(9): 553–564.

Hosterman, Heather R., McCornick, Peter G., Kistin, Elizabeth J., Sharma, Bharat, and Bharati, Luna, 2012, 'Freshwater, climate change and adaptation in the Ganges River basin', *Water Policy*, 14(1): 67–79.

Immerzeel, W. W., Van Beek, L.P.H., and Bierkens, M.F.P., 2010, 'Climate change will affect the Asian water towers', *Science,* 328(5984): 1382.

Institute for Defence Studies and Analyses, 2010, *A Task Force Report-Water Security for India: The External Dynamics.* New Delhi: IDSA: 56–57.

Jianchu, X., Shrestha, A., Vaidya, R., Eriksson, M., and Hewitt, K., 2007, *The Melting Himalayas: Regional Challenges and Local Impacts of Climate Change on Mountain Ecosystems and Livelihoods*, ICIMOD Technical Paper, June 2007.

Kulkarni, A. V., Bahuguna, I. M., Rathore, B. P., Singh, S. K., Randhawa, S. S., Sood, R. K., and Dhar, S., 2007, 'Glacial retreat in Himalaya using Indian Remote Sensing satellite data', *Current Science*, 92(1).

Mallet, Victor, 13 October 2013, 'After the flood: climate change blamed for Himalayan tragedy', *Financial Times.*

Mehra, Malini, 2009, 'India starts to take on climate change', *Climate Connections State of the World 2009.* Washington, DC: The Worldwatch Institute.

Milly, P.C.D., Dunne, K. A., and Vecchia, A. V., 2005, 'Global pattern of trends in streamflow and water availability in a changing climate', *Nature*, (438): 347–350.

Mirza, M.M.Q., 1997, 'Hydrological changes in the Ganges system in Bangladesh in the post-Farakka period', *Hydrological Sciences Journal*, 42(5): 613–631.

Mirza, M.M.Q., 2011, 'Climate change, flooding in South Asia and implications', *Regional Environmental* Change, 11(1): 95–107.

Mirza, M.M.Q., Warrick, R. A., and Ericksen, N. J., 2003, 'The implications of climate change on floods of the Ganges, Brahmaputra and Meghna Rivers in Bangladesh', *Climatic Change,* 57(3): 287–318.

Morton, K., 2011, 'Climate change and security at the third pole', *Survival,* 53(1): 121–32.

National Intelligence Council, 2009, *India: The Impact of Climate Change to 2030.* The National Intelligence Council Commissioned Research Report.

National Water Mission (NWM), 2009, 'Ministry of water resources, government of India'. Available at: http://india.gov.in/allimpfrms/alldocs/15658.pdf.

OutlookIndia.com, 2 September 2011, 'B'desh to propose joint basin management to PM', accessed 16 February 2015 at http://www.outlookindia.com/news/article/Bdesh-to-Propose-Joint-Basin-Management-to-PM/733401

Palmer, M. A., Liermann, R., C. A., Nilsson, C., Flörke, M., Alcamo, J., Lake, P.S., and Bond, N., 2008, 'Climate change and the world's river basins: anticipating management options', *Frontiers in Ecology and the Environment,* 6(2): 81–89.

Pandey, C. L., 2011, 'Climate change policy in Nepal', *Nepalnews.com*, 8 December 2011.

Podesta, J., and Ogden, P., 2008, 'The security implications of climate change', *Washington Quarterly,* 31(1): 115–138.

Program Support Unit of the Agricultural Sector Program Support, 2009, Phase II, *FINAL Climate Management Plan.* Agricultural Sector: Bangladesh.

Salman, S.M.A., and Uprety, K., 1999, 'Hydro-politics in South Asia: a comparative analysis of the Mahakali and the Ganges treaties', *Natural Resources Journal,* 39, (2): 295–343.

Schellnhuber, H. J., Hare, B., Serdeczny, O., Schaeffer, M., Adams, S., Baarsch, F., and Rocha, M., 2013, *Turn Down the Heat: Climate Extremes, Regional Impacts, and the Case for Resilience.* Washington, DC: World Bank.

Sharma, Bharat R., and Sharma, Devesh, 2008, *Impact of Climate Change on Water Resources and Glacier Melt and Potential Adaptations for Indian Agriculture.* New Delhi: International Water Management Institute.

Singh, P., and Arora, M., 2007, 'Water resources potential of Himalayas and possible impact of climate', *Hydrology Review*, 22: 109–132.

Srabani, Roy, 1 December 2010, 'Climate change and water sharing in South Asia: conflict or cooperation?' The Asia Foundation, accessed 20 March 2014 at http://asiafoundation.org/in-asia/2010/12/01/climate-change-and-water-sharing-in-south-asia-conflict-or-cooperation/

Suzanne Goldenberg 22. April 2009. Climate change threatens Ganges, Niger and other mighty rivers. The Guardian. Retrieved 1 March 2015. www.theguardian.com/environment/2009/apr/22/drought-environment-waterways

Swain, A., 2004, *Managing Water Conflict: Asia, Africa and the Middle East.* London: Routledge.

Swain, A., 2010a, 'Environment and conflict in South Asia: water-sharing between Bangladesh and India', *South Asian Journal*, 28: 27–34.

Swain, A., 2010b, *Struggle Against the State: Social Network and Protest Movements in India.* Aldershot: Ashgate.

Swain, A., 2011, 'South Asia, its environment and regional institutions', in Lorraine Elliott and Shaun Breslin (eds.), *Comparative Environmental Regionalism.* London: Routledge: 76–91.

Swain, A., 2012, 'Global climate change and challenges for international river agreements', *International Journal on Sustainable Society,* 4(1–2): 72–87.

Times of India, 18 June 2013.

Upreti, B.C., 1993, *Politics of Himalayan River Waters: An Analysis of the River Water Issues of Nepal, India and Bangladesh.* Jaipur: Nirala.

Verghese, B. G., and Iyer, Ramaswamy R., 1993, *Harnessing the Himalayan Rivers: Regional Cooperation in South Asia.* New Delhi: Konark Publishers.

Watts, Jonathan, 2010, 'Chinese engineers propose world's biggest hydro-electric project in Tibet', *The Guardian*, 24 May 2010.

Wax, Emily, 17 June 2007, 'A sacred river endangered by global warming', *The Washington Post.*

Webster, P. J., Jian, J., Hopson, T. M., Hoyos, C. D., Agudelo, P. A., Chang, H. R, and Subbiah, A., 2010, 'Extended-range probabilistic forecasts of Ganges and Brahmaputra floods in Bangladesh', *Bulletin of the American Meteorological Society*, 91(11): 1493–514.

Wirsing, R. G., and Jasparro, C., 2007, 'River rivalry: water disputes, resource insecurity and diplomatic deadlock in South Asia', *Water Policy*, 9: 231–251.

Wong, Edward, 1 June 2011, 'Plans for China's water crisis spurs concern', *The New York Times.*

World Bank, 2010, *South Asia Water Initiative (SAWI) Multi-Donor Trust Fund Annual Report FY10 (July 2009 – June 2010),* Prepared by the World Bank for the 3rd Annual Donors Meeting Dhulikhel, Nepal 28–29, September 2010.

Xu, J., Grumbine, R. E., Shrestha, A., Eriksson, M., Yang, X., Wang, Y., and Wilkes, A., 2009, 'The melting Himalayas: Cascading effects of climate change on water, biodiversity, and livelihoods', *Conservation Biology*, 23(3): 520–30.

Yao, Tandong, Thompson, Lonnie, Yang, Wei, Yu, Wusheng, Gao, Yang, Guo, Xuejun, . . . Joswiak, Daniel, 2012, 'Different glacier status with atmospheric circulations in Tibetan Plateau and surroundings', *Nature Climate Change*, 2: 663–667.

3 The Jordan River basin

Background to the basin

The Jordan basin is at the heart of the water dispute in the Arab–Israeli water conflict. The protracted conflict in the regions directly affects the prospects for joint management of shared water resources, in some ways similar to how the discourse on the Ganges (also discussed in this volume) is intimately linked to the political relations in the basin. The river has witnessed many attempts at establishing joint and effective transboundary water management (TWM), owing to its political importance in this dry and conflict-ridden region (Jägerskog 2003; Zeitoun 2006). The Jordan basin covers an area of close to 18,000 km² and includes parts of Lebanon, Syria, Jordan, Israel, and Palestine. The sources of the Upper Jordan River are primarily three major springs – the Hasbani, with its source in Lebanon the Dan, with its source in Israel, and the Banias, with its source in the Golan Heights, which is occupied by Israel from Syria. The river flows south and drains into Lake Tiberias in Israel before it moves southward towards the Dead Sea. However, south of Lake Tiberias, the Jordan River is dammed, and the only water that leaves Lake Tiberias is the water that is pumped into the National Water Carrier of Israel (FoEME 2012). In mid-May 2013, Israel started to release a small portion (1,000 m³ per hour) from Lake Tiberias into the Jordan River (Rinat 2013), which is a good step but by no means a major change in the management of the river. The tributary Yarmouk River, with its source in Syria, joins the Jordan River from the east after having flowed through Jordan. The Yarmouk is the main source of water for Jordan and a large part of the water flow from the Yarmouk is diverted upstream in Syria, then later in Jordan, then into the King Abdallah Canal before joining the Jordan River (Haddadin 2001). Currently, inflow levels to the Dead Sea from the Jordan River are around 5 per cent of the original inflow levels, due to diversions upstream of the Dead Sea. This leads to massive environmental problems, such as proliferation of sink holes at the former bottom of the Dead Sea, as well as a shrinking of the Dead Sea water level by around one metre annually. Already, the Dead Sea has lost one-third of its original surface area (FoEME 2012). For Israel and Jordan, the Jordan River basin (which includes its tributaries) is of immense importance as both states withdraw large percentages of their water from it. For the other riparian states in the basin – Syria and Lebanon – the Jordan River basin is not as important as it is for Israel, Jordan, and (potentially) the Palestinian areas. Lebanon gets the greater part of its water from the Litani and Awali Rivers, and Syria receives most of its water from the Euphrates and the Orontes. Moreover, the quantity of water in the Jordan River is constantly declining along the course of the river as more water is withdrawn from it than is renewed each year. This is especially significant in years of drought with detrimental effects on the Dead Sea, as noted above. The decline in the flow threatens the quality of the water, as saline water can infiltrate, making it impossible to use. Agricultural drainage water, untreated waste water, and water from saline springs are draining into the river from both sides. This is accelerating the deterioration of the water quality. The challenges relate both to quality as well as quantity of the water (Wolf & Hamner 2000). As can be noted in the Table 3.1, which shows water availability in the

Table 3.1 Water availability in the MENA region[1]

Country	Total internal renewable water resources (km³/year)	Groundwater: produced internally (km³/year)	Surface water: produced internally (km³/year)	Water resources: total renewable per capita (m³/capita year)
Turkey	227.00	69.00	186.00	2,889
Iraq	35.20	3.20	34.00	2,652F
Iran	128	49.30	97.30	1,957
Lebanon	4.80	3.20	4.10	1,110
Morocco	29.00	10.00	22.00	940F
Syria	7.13	4.84	4.29	865
Egypt	1.80	1.30	0.50	773
Tunisia	4.20	1.50	3.10	450
Djibouti	0.30	0.015	0.30	367F
Algeria	11.20	1.49	9.76	350
Oman	1.40	1.30	1.05	550F
Israel	0.75	0.50	0.25	261F
Jordan	0.682	0.45	0.485	164F
Bahrain	0.004	0.00	0.004	157
Libya	0.60	0.50	0.20	99.4F
Saudi Arabia	2.40	2.20	2.20	99.3F
Yemen	2.10	1.50	2.00	96.6F
Palestine	0.812	0.74	0.072	94.7*
Qatar	0.056	0.056	0.00	70.6
United Arab Emirates	0.15	0.12	0.15	35.3
Kuwait	0.00	0.00	0.00	7.20

F – FAO estimate.

* Data from Palestinian Water Authority 2007. Includes approximately 53.4 thousand million cubic feet (cmc) purchased externally from an Israeli company. If the externally purchased water is excluded, the total renewable water resources per capita and year for Palestine would be approximately 79.6m³.[2]

Middle East and North Africa (MENA) region at large, all the riparians in the Jordan basin have a water scarcity situation – particularly Israel, Jordan, and Palestine.

There have been many attempts at promoting cooperation in the Jordan basin, dating almost right back to the establishment of Israel in 1948. During the 1950s, an American-led plan known as the Johnston-plan (named after the US envoy Eric Johnston) was developed, suggesting a division of the waters of the Jordan. While viewed relatively positively in the basins, it could not be adopted for political reasons (as that would mean a *de facto* recognition of Israel by its Arab neighbours). Prior to that, there had been unilateral developments in both Israel and Jordan (but also from Syria and Lebanon) focusing on harnessing the water of the Jordan River (Jägerskog 2003). While there was no formal agreement on the Johnston plan, there emerged a tacit understanding (with the help of US financial assistance) between Israel and Jordan, who developed, respectively, the National Water Carrier (Israel) and the East Ghor Canal (Jordan), which contributed to the draining of the Jordan River (Shapland 1997). During that period, and under UN auspices, discussions relating to coordination on the Jordan River emerged and were upheld on a regular basis between Israel and Jordan – the so-called 'Picnic Table Talks', which functioned up until the signing of the Israeli–Jordanian Peace Agreement in 1994 (Wolf & Hamner 2000).

Between Israel and Palestine, the thrust of the conflict is focused on the shared aquifers. Much of the rain that falls over the West Bank makes its way into aquifers that are then 'flowing' into Israel. When Israel occupied the West Bank in 1967, it gained control over the sources of the water it shares with the Palestinians and has largely dominated the discussions and practice in their use since then (Jägerskog 2008; Selby 2013).

The Israeli-Palestinian agreement resulted from negotiations that followed the signing of the Declaration of Principles (DOP) from September 1993. Those negotiations were made possible largely by the secret talks being held in Oslo, but were also a follow-up to the bilateral talks that followed the Madrid Conference in October 1991. The DOP identified water as one of the key issues that needed further negotiations. The DOP, which made possible more open and official negotiations between Israel and the Palestinians, was sparse on details relating to water, but the 1995 interim agreement provided more detail. It clarified that the Palestinians had water rights (however, not outlining what those rights consisted of). The water issue was regarded as one of the difficult issues between the parties, and its final solution was postponed to the permanent status negotiations. The parties agreed to coordinate the use of shared waters in the interim period, for which a Joint Water Committee (JWC) was established. The agreement in principle maintains prior allocations to Israel, but also outlines increased Palestinian allocations – however, they are from largely undefined sources. The agreement is ambiguous in terms of how these sources should be developed and how their finances should be handled. The intention was that the final status negotiations between the parties should have been concluded by 1998. However, such an agreement has yet to materialise (Jägerskog 2003). The arrangement between Israel and Palestine is naturally (since it is only a framework agreement waiting for a final one) not comprehensive in its scope, and as Selby (2013) notes, the (mal-)functioning of the JWC has deprived the Palestinians effectively of the chance to develop their own water sector.

The Israeli–Jordanian Peace Agreement from 1994 was in some respects a natural follow-up to the Israeli-Palestinian DOP. Israel and Jordan had for relatively long viewed it in their long-term interest to reach a peace agreement. However, King Hussein of Jordan could not go further politically in rapprochement with Israel than the Palestinian leadership. Once they had agreed on the DOP, Jordan was able to negotiate with Israel. Israel and Jordan had for some time enjoyed a tacit understanding on issues such as water, which were formalised in the Peace agreement in 1994. Water is dealt with rather extensively in the agreement. The agreement outlines the water allocations to which both countries are entitled from the Jordan River, its tributary Yarmouk, as well as the shared groundwater in Wadi Araba/Arava. The agreement is interesting in that it also provides for the storing of Jordanian 'winter water', when they do have a relative surplus of water flow, in Lake Tiberias, inside of Israel, which subsequently releases the water in the dry summer period when Jordan needs it in urban centres such as Amman.

The approaches used in efforts to reach a settlement of the water issue are different in the two cases. While the Israeli–Palestinian agreement came out of a protracted conflict that had been going on for years and where informal forms of cooperation were largely absent, the Israeli–Jordanian agreement was to some extent an agreement that codified some of the existing practices when it came to issues, while allowing the issues to develop further. Israel and Jordan had, in spite of being formally in conflict, unofficially met regularly to discuss and coordinate action on the Jordan River. Unofficially, there had been many 'track-two' efforts between Israeli and Palestinian academics to discuss the water sharing between the parties. This, to some extent, provided some basic understanding of the issues between the parties. However, there were many issues that complicated the water part of Oslo II, primarily that water related to many other sensitive issues destined for the final negotiations such as borders, settlements, and the Palestinian refugees' status (Jägerskog 2003). The efforts for a final agreement led by the Obama Administration under the lead of Secretary of State Kerry (2013–2014) have largely been kept out of the public's eye. Thus, not much has been reported about water (or its relation to climate change), but clearly water is part of the discussions, albeit not the most important one.[3]

The advantage in the case of the Israeli–Jordanian agreement was that there existed a code of practice, and the issues were understood by the two sides, which provided for easier negotiations. The ambition was also to try to solve most of the water issues directly in the agreement. Admittedly, some ambiguity is present, in particular in relation to allocations in times of less water available in the Jordan river system to cover allocation (not unheard of in a region which usually experiences a drought every seventh year), but most issues are dealt with in detail.

The approach in the Israeli–Palestinian case was to adopt a 'step-by-step' approach by which the parties agree on certain issues while postponing some harder issues. Thereby, there is an inbuilt ambiguity in the agreement, which may be what is needed for the parties to agree. However, it carried the downside that in the end it is not possible to agree; i.e. the conflict is in effect postponed, and in some cases worsened.

While the dispute between Israel and Jordan mainly relates to the Jordan River, the water dispute between Israel and Palestine primarily relates to their shared aquifers. In the West Bank, the mountain aquifer is the source of contention. The aquifer is divided in the western aquifer, which flows from the highest parts of the West Bank westwards, and the northeastern aquifer, which flows northeast into Israeli territory and the eastern aquifer that flows east towards the Jordan River. The eastern aquifer is not considered to be a transboundary water resource, as its flow is almost entirely within the West Bank (Assaf et al. 1993). These aquifers are recharged through the rainfall over the West Bank. As the rainfall varies from year to year, the recharge is subject to major variations from year to year. Here, the climate variability, manifested in rainfall variation, evidently complicates the relations among the riparians, and this is particularly evident in years of drought. It has been noted by many researchers that the variations in water availability are possible to account for (Kolars 2000; Fischhendler 2004; Falkenmark & Jägerskog 2010; Cooley & Gleick 2011). If parties agree to negotiate allocations on the basis of 'reliable' water, and include provisions for the allocation of the 'non-reliable' extra water, the problem of allocation in times of drought would be more easily dealt with. However, states tend not to be too keen to adopt flexible agreements that will not provide them with well-defined cubic metres of water. Which, of course, is an illusion, and sometimes an instigator of conflict, as there nevertheless will be a certain amount of water in the river or aquifer any given year (Falkenmark & Jägerskog 2010). Kolars (2000) and Fischhendler (2004) have noted that the Jordan River basin is subject to high seasonal and multi-annual variances in precipitation and consequently flow in the river or water available in the aquifers. Jägerskog (2003) noted that there are no provisions in the present agreements to allow for flexibility.

A further source of dispute between Israel and Palestine is the Israeli coastal aquifer, which runs along the Mediterranean coast in Israel and connects from Israel into the Gaza aquifer, which underlies the Gaza Strip. There exists general agreement that the Gaza aquifer is an extension of the coastal aquifer in Israel, although there are different views on the extent to which they are connected. However, there is a general agreement that the flow in the aquifer is predominantly east–west, which seems to indicate that Israeli activities north of the Gaza Strip will not affect the part of the aquifer beneath the Gaza Strip very much, nor will activities in the Gaza strip affect the Israeli coastal aquifer very much (Shapland 1997).

As noted above, the Jordan basin is characterised by relatively large climate variability. Dry years are followed by wetter (in relative terms) years. This has been the case all through history and has been a feature to which managers of water, such as farmers, have had to accommodate their ways of managing water.

Climate change predictions for the Jordan basin are complex, and estimations vary. However, most prediction point to decreased rainfall. According to a climate change vulnerability assessment by UNEP (2012), it is clarified that climate change will put additional pressure on the water resources in that variability will increase and predictability of rainfall (already low) will decrease. IPCC has reported that 'annual rainfall is likely to decrease in much of Mediterranean Africa and Northern Sahara, with the likelihood of a decrease in rainfall increasing as the Mediterranean coast is approached' (IPCC 2007: 866). The Levant (Israel, Palestine, Syria, and Lebanon) is by many deemed to be the part of the MENA region, most affected in terms of precipitation changes but, coupled to that, also an increase in average temperatures increasing the pressure being felt (Alpert et al. 2008). Averages for the MENA region suggest between 10–30 per cent in decrease in rainfall during the 21st century, which is very significant (Sowers et al. 2011).

The Jordan River is small by comparison to almost all other internationally contentious rivers, but since it is one of the few available water sources in the region, its importance is great. In terms of delimitations, this chapter will address the Jordan basin, but not in its entirety. It will focus primarily on the Lower Jordan basin and the groundwater sources pertaining to Israel, Jordan, and Palestine. Syria and Lebanon are not addressed in this chapter, as their interactions with the other basin states (in particular Israel) are minimal. Syria, Israel, and Lebanon do not have formal relations with each other. Thus, discussions on their shared waters (and related climate change) are minimal.

Institutional arrangements

As previously noted, water is highly political in the Jordan basin, both on a domestic level as well as between the countries sharing the waters of the basin (Allan 2001). Indeed, water was identified as one of the five key areas that should be negotiated in the 'final status' negotiations between Israel and the Palestinians. Water was also identified as a key area in the multilateral track that was formed after the Madrid Peace conference in 1991. Ever since the State of Israel was proclaimed in 1948, a number of initiatives have been taken regarding the utilisation of water resources. The issue of water is part of the peace treaty of 1994 between Israel and Jordan, of the Declaration of Principles between Israel and PLO in 1993, and of the Oslo II agreement from 1995. Still, no agreement that includes Lebanon and Syria has been reached, for political reasons. While analysts previously argued that water may be a source of war in the region, research has shown that in spite of the political instability in the Middle East, cooperation over water has functioned and continues in different forms (Jägerskog 2003). Israel and Jordan coordinated water issues as early as the 1950s, in spite of the fact that there was no formal peace (Haddadin 2001). Water experts from countries in the region have continually met to conduct a dialogue on issues connected with water – there is, among others, a group called EXACT,[4] which has representatives from Israel, the Palestinian Authority, and Jordan meet with a number of development partners to discuss common water projects.

Even though cooperation is important, it will not alleviate water scarcity. One important factor that is neglected by those who are convinced that wars about water are likely in this region is that virtual water is exported from certain regions. Through the import of virtual water to Jordan, Israel, Jordan, and Palestine, water that would normally be needed for agriculture is available in sectors such as industry and households. In order to manage the increasing population in the region and its need for food, greater imports of water-rich crops will be required (Allan 2001). While Israel can cope with such an increased import of food economically, the situation, primarily for the Palestinians but also for Jordan, is more difficult, both economically and, most importantly, socially. They would have to manufacture other products for the export market in order to finance the import of increasing quantities of virtual water. In Palestine in particular, it is imperative to find alternative employment for many of those employed in the agricultural sector. In Gaza there are plans to build a new desalination plant to cater to the needs of the continuously growing population there.

The fact that agreements on water exist does not mean that the issue is resolved, but remains one of the final status issues to be negotiated between Israel and the Palestinian Authority. At present there is no agreement on how the water resources should be divided. But the agreements have provided for the establishment of a Joint Water Committee (JWC), which is in charge of coordinating water issues. Rather predictably, the Israeli–Palestinian arrangements to solve the conflict by a 'step-by-step' approach have encountered a range of challenges. There has been coordination in the JWC, but since the JWC deals with only projects in the West Bank – and is based on consensus decision making – it in effect provides Israel with a veto over Palestinian projects. This is something that has aggravated Palestinian frustration and sense of inability to develop their water sector. While such an arrangement would have been acceptable until 1998 – when the final status negotiations were supposed to be finalised – they have not served the parties (primarily not the Palestinians) very well during the last decade. The JWC has, however, managed to keep operating throughout the conflict – even during the Palestinian uprisings – but its effectiveness is severely hampered (Jägerskog 2003, 2008).

The Israeli–Jordanian agreement has been effective. In a region plagued by conflict, the fact that the treaty has been abided by is noteworthy. The storing arrangements of 'Jordanian' water in Lake Tiberias in Israel and Israel's subsequent release has functioned, and that in itself has been a trust-building feature of the arrangements (Jägerskog 2003). There have been disagreements relating to how to share the deficiency in drought years, but the matter has been solved through their JWC.

The efforts through the EXACT and the two Joint Water Committees are at best described as embryonic attempts at institutionalisation of cooperation on the waters in the Jordan basin. In the case of Israel and Jordan, there is a peace agreement that governs their water relations, but the level of institutional mechanisms beyond the JWC is limited. In the case of Israel and Palestine, there does not exist a final agreement that deals comprehensively with the water resources that they share. The arrangements in the region say nothing of the (in many cases) sophisticated structures of the River Basin Organizations governing the other transboundary basins analysed in this volume. Neither the institutional structures nor the inclusiveness of the basin are evident, as Lebanon and Syria are not part of the basic basin structure. If agreements exist at all, they are all on the bilateral level. Any effort to discuss climate variability and change on a basin-wide level are not possible at present and discussions of the issue at bilateral level are sparse. De Stefano et al. (2010), in their global study of treaty institutions on transboundary waters, noted that the ability of river basin organisations and the like, and the mandate and institutional capacity of these institutions' were critical for the ability to manage conflicts or country grievances, something that is likely to rise with climate change and increased climate variability.

Even though Israelis and Palestinians cooperate on water, this cooperation is not without problems. The power asymmetries between the parties are important components for understanding the outcomes of the transboundary water interaction and the quality and 'fairness' of the cooperation (Zeitoun & Mirumachi 2008). There are clear differences in perspectives between the parties concerning the way future water needs are to be managed. In the case of Israel and the Palestinian Authority, both parties agree that a basic level of cooperation and coordination is essential, but the views of content of the cooperation are not the same. Analysing the cooperation in the JWC shows the 'domination' by Israel (Selby 2003). For a decision to be made in the JWC, a consensus has to be reached. Since all decisions concern projects in the Palestinian areas, this means that Israel holds a veto for Palestinian projects. The power that Israel holds in the JWC makes it possible for them to dominate the Palestinians (Zeitoun 2008). The World Bank report from 2009 on obstacles to Palestinian development of its water sector describes in detail how such power is used at the Israeli–Palestinian JWC, and the fact that a project has been approved by the committee is not a guarantee for it to actually become implemented (World Bank 2009; Selby 2013). The civil administration in Israel (in practice the Israel Military Defence Forces, IDF) has the right to stop water projects for security reasons. This right is frequently used. Palestinian representatives needing to provide their people with water are frustrated over the fact that they are not able to implement projects in Palestinian areas. Israel, on the other hand, argues that they cannot compromise their security. It is worth noticing that these conflicts are not necessarily violent, but can be rather subtle and take place underneath a facade of cooperation. It may be Israel preventing Palestinian projects in the JWC, or constructing the wall/safety barrier on top of significant springs, etc. (Zeitoun 2008).

Actors and interests

As noted by Messerschmid (2012), many development partners (Germany and UNDP, for example) are interested in the Jordan basin as well as in the issue of climate change and have projects and programmes directed towards this area. In the case of Palestine, UNDP has engaged in developing an adaptation strategy with a number of no-regret and low-regret options that would allow them to address some of the climate change challenges that they are facing (UNDP PAPP 2010). Messerschmid (2012) as well as Zeitoun and Jägerskog (2009) note that in many cases, development partners active in the basin, particularly engaged on the Palestinian side, do primarily view climate change (or engaging at the transboundary water level) as a technical engagement whereby adaption measures should be identified, but this seems largely to be the case in Jordan as well. Thus, often a political analysis of the proposed adaptation measures is not part of the preparatory work by the partners engaging, increasing the risk for not accounting for how climate change and water may become political tools in the discussions over water sharing and allocation.

GCC entering the basin

Focus on climate change in the Jordan basin has, during the last few years, increasingly permeated the discourse in the region. Since the Middle East in general, and perhaps the countries that are part of the Jordan basin in particular, are characterised by conflicts and tensions, it comes as no surprise that the potential effects of climate change represent an area of concern (Trondalen 2009; Feitelson et al. 2012). Not least since one of the potential effects of climate change is a decrease in the water availability, something that would be a major blow in an already water-scarce region. It has been noted that in the region the agreements that exist are not suited to deal with the climate variability (the effect that water availability often varies significantly between years) in the region. The agreement between Israel and Jordan and the interim agreement between Israel and the Palestinians Authority relating to water include fixed allocations, which are based on multi-year averages and thereby lack flexibility (Falkenmark & Jägerskog 2010).

At the NGO level, the Friends of the Earth Middle East has highlighted the implications of climate change in the region and its implications on the water sector(s) in the basin (FoEME 2007). They identify climate change as a 'threat multiplier' that acerbates already-challenging situations, not least relating to water, and with the poor and vulnerable facing the greatest risk. While the work of NGOs such as FoEME is well developed and well researched, it still has not gained significant attention in the region. The GLOWA Jordan project is another initiative to study the effects of climate change in the Jordan basin. This project is a largely scientific exercise taking a multi-disciplinary approach. It analyses natural scientific projections and results as well as socio-economic factors relating to the water aspects of climate change in the basin.[5]

The prevalent impression from discussions in the basins with official representatives, academics, and NGO representatives is that the climate change discussions at the global level are receiving some attention, but primarily by the elites. In Israel, Jordan, and Palestine, the focus is primarily elsewhere – in Israel, the focus is on improvements of the water reuse and desalination, which they view as necessary in any case. In Palestine, climate change is primarily addressed through projects and not as an overarching programme for the country (Mimi, personal communication 2012). In Jordan, the view that is emerging is that climate change is evident in the discourse but is being picked primarily by the elites. It seems that the predominant narrative is that climate change will add another burden on the already water-stressed country. In Palestine, the predominant sense is that they are in a 'crisis management mode', which lends little time and resources to address the issue of climate change. There are many more-pressing issues, as seen from the Palestinian perspective, and their main focus is (apart from ending occupation and gaining water rights) to rehabilitate non-functioning wells and drill new ones (this process is slow, however, since all Palestinian projects need to go through the JWC, where Israel can veto all projects). Nevertheless, through support for climate change adaptation projects by development partners in the region, the Palestinians and the Jordanians are starting to engage with the concept and its practical consequences for water. However, as noted by Messerschmid (2012), the discourse in the basin, which for many years has been one of securitisation (Jägerskog 2008), has added climate change to the securitisation framework, thereby providing the actors further reasons for hard positions with regards to their transboundary water resources.

At the level of the media in the basin, the reporting varies. However, much of the reporting relays the message that lack of rain and the following droughts and water scarcity is not a direct consequence of climate change, but rather of normal climate variability. This seems to be the case both in English press as well as in the Arabic press (Beck & Collet 2011). Furthermore it is worth noting that the increasing reach of regional satellite networks such as Al-Jazeera and Al-Arabiya, which have hosted both documentaries as well as talk shows on the topic of climate change, has placed general knowledge of climate change issues higher up on the agenda in the MENA region.

In Jordan, however, the media has been reporting the lack of rain and sub-sequent water scarcity was not a result of climate change, but rather of natural variations in the system, something that has been prevalent for a long time (Oamri 2011). In Israel, the media has been reporting prolonged droughts, sometimes with connection to climate change and sometimes without. In any case, the reporting has allowed the Israeli Water Authority to apply some demand management measures in the country (Sowers et al. 2011; Aviram, personal communication 2012). In Palestine, the media has been largely focused on the water crisis and the ways to overcome that situation. Climate change is not a major focus of the media; the reporting focuses more on the unequal water situation and the restrictions imposed by Israel on the Palestinian water sector (Messerschmid 2012).

Israel

The Israeli water geography is rapidly changing. The advent of large-scale desalination, as well as increased efficiency in terms of water for agriculture (increase of re-used water), is adding to the Israeli water budget (Feitelson & Rosenthal

2011). The Israeli Water Authority estimates that by 2020, around 40 per cent of the water budget will be covered by desalinated water (around 750 mcm out of a total of 1,950 mcm) and on top of that an additional 528 mcm of recycled wastewater will be added to the agricultural water budget (IWA 2010 in Feitelson & Rosenthal 2011).

The fact that the Israeli water budget is increasing at a quite rapid pace puts Israel in a relatively good position to manage demands coming from increasing climate variability as well as from climate change (Feitelson, personal communication, 2012). In addition, the relatively recent discovery of natural gas in Israel also brings down the energy costs for the desalination, thus making it further plausible as a more long-term financially relative sustainable option (Hammar-skjöld, personal communication 2012). As has been noted, the introduction of large-scale desalination also permits emergence of positive-sum outcomes at the transboundary level, where all parties can gain (Phillips et al. 2007).

Ram Aviram, the former Israeli head negotiator of the Multilateral Peace Talks, notes that the desalination is a 'game-changer' in that it significantly increases the Israeli water budget and thereby enables potentially more compromise on their transboundary waters (Aviram, personal communication 2012). However, such a development is not a given and is much doubted by Arab analysts, such as El Khoury (2012), who maintains that Israel will not change its policy unless it has to.

Jordan

According to AFED, the general understanding of climate change in Jordan is relatively high. It is higher than in other Arab countries (AFED 2009). While the AFED conclusions had methodological shortcomings (using readers of *Al-Doustor*, a major newspaper, and inviting people who read it to respond meant that those interested in the issue were more likely to respond), it seems to at least indicate that the understanding of climate change, its underlying reasons, as well as potential implications are relatively well understood in Jordan – at least in some parts of society.

Also, in Jordan, the water situation is changing. In the aftermath of the war in Iraq, many Iraqi refugees and migrants entered Jordan (primarily Amman but also other areas), which put additional pressure on their water resources, and the ongoing Syrian crisis with the influx of refugees puts yet additional pressure on the already-challenging situation. This may imply further re-allocation of water away from the agricultural sector to domestic and industrial uses, something that is politically costly in a country like Jordan. Jordan, before considering climate change, was already experiencing water shortages. The additional burden increases the importance of increasing supply-sided solutions, such as the pumping of the Disi aquifer in the south of Jordan (which is shared with Saudi Arabia), as well as their focus on the Red-Dead project, which would bring water from the Red Sea to the Dead Sea but also desalinate some of the water and provide it in the Jordan national water system, thus increasing its water budget. The Jordanian Water Strategy for the period of 2008–2022 plans to increase the desalination part of Jordan's water budget from 1–31 per cent by 2022. Similarly, it plans to extend the use of recycled waste water from 10–13 per cent, while decreasing groundwater pumping from 32–17 per cent of the water budget (Jordan Water Strategy 2008). Jordan also adopted its first climate change policy (being the first Arab country to do so) in May 2013. The policy is for the period of 2013–2020 and it takes stock of UNDP-supported studies on the predicted effects of climate change in Jordan, which include increased frequency of flash floods, intense rain, and droughts (Namrouqa 2013).

The efforts being planned for Jordan will enable them to better meet the increased climate variability, which would be a result of climate change. However, high population growth in Jordan up until 2010 creates a situation where the water scarcity increases. This is explained by a high birth rate, but also by a large influx of refugees – first in 2003 and 2004 from Iraq, then during 2012 from Syria.

Palestine

The Palestinian water sector is primarily focused on 'crisis management' as described by the Palestinian Minister for Water, Shaddad Attili (personal communication 2012). Thus the attention that is being paid to issues such as climate change is not great. Other political priorities and urgent issues take precedence over climate change discussions and debate. This is reiterated by Mimi (personal communication 2012), who underlines that climate change is not a political priority of the Palestinian Water Authority (PWA), while the Environmental Quality Authority (EQA) is more engaged in it. However, the UNDP PAPP report (2010) suggests that all the relevant ministries and agencies are all for planning for adaptation measures as suggested in the UNDP PAPP report, which recommends no-regret measures such as development of flood contingency plans, establishment of clear water-use priorities, increased (sustainable) production of freshwater, and increased use of brackish and re-used treated waste-water. At the nongovernmental level, however, the support for adaptation seems more qualified on the ability to place the adaptation measures into its proper political context (UNDP PAPP 2010).

The PWA is, besides managing the 'constant crisis', also engaged in work to reform the Palestinian Water Sector. With donor support, the PWA is working to clarify its role as a regulator of the water sector in Palestine, rather than as a project implementing agency. Furthermore, delimitations of the mandate of PWA in relation to the Ministry of Agriculture is being discussed. To address climate change, the water sector in Palestine needs to address management issues but is severely restricted since Israel in essence controls the resources, partly through the consensus procedure of the JWC but also through the civil administration in the West Bank (Selby 2003; Jägerskog 2008). In addition Palestine also has a high population growth rate, in particular in Gaza but also in the West Bank.

Responses in/from the basin – adapt, resist, subvert

As noted above, we are aiming at analysing the TWM arrangements – as well as the different stakeholders involved and influenced by these arrangements – and thereby draw conclusions on how the climate change discussion has been adopted and/or adapted to serve the political or socio-economic interests of the several actors in each basin. As is evident from the above presentation of the Jordan basin situation, when there is a level of ambiguity, as is the case with regards to climate change in the Jordan basin (UNDP PAPP 2010), it can be used as well as manipulated by different actors with different objectives.

In the Jordan basin there is evidence of all the categories outlined as components for our analytical framework. The picture that emerges is one of a certain degree of adaptation measures happening in the basin, while simultaneously there is a degree of resistance to include it in management frameworks. On top of that, the climate change argument is being used as a political means to some degree, i.e. a subversion of the discussion. Below is a description of various categories in which adaptation, resistance, and subversion is happening.

Adapt

Organisational

In all countries, increased funding has been made available to adapt to the effects of climate change. In some cases, climate change (or the anticipation of it coming) was not what led to the financial adaptation measures. There is a focus from the MENA countries at large – and also in Jordan – to focus on supply-sided solutions, rather than focusing on managing demand, although such perspectives are also gaining ground. Increased use of fossil water, desalination, and increased re-use of water are important strategies (Sowers et al. 2011). At the regional level, there is no organisation dealing with climate change that includes all parties. The UN Economic and Social Commission for West Asia (UNESCWA) works on climate change and water but that does not include Israel in the basin and their focus is less on transboundary water relations.[6]

System

It is clear that in comparison to most of the other cases in the volume, the systems that adapt to climate change are less developed at the regional scale. This is of course due to many of the protracted political challenges affecting the region. Still, through the UNESCWA work, Palestine and Jordan are involved in efforts aimed at establishing more systems to provide early warning for prolonged droughts, etc. It should be noted, however, that the region is used to droughts with relative frequency, which has created a relative sense of preparedness. Most of the efforts to address challenges are related to structural initiatives (see below).

Structure

In Israel, the investment in desalination plants is something that has been going on since some 10–15 years but has accelerated during the last five years (Feit-elson & Rosenthal 2011; Aviram 2012). Once finalised, the desalination will be supplying 25–30 per cent, if not more, of Israel's annual water budget. While this assists Israel in adapting to climate change and increased variability, its prime motive is arguably not primarily caused by climate change but rather to address the water scarcity situation in the country in general. It certainly helps Israel in managing the recurrent droughts (climate variability) that hit the region. Furthermore, the increased focus on re-use of agricultural products is also something that will aid adaptation, but also, in this case, the prime motivation is rather the water scarcity situation in general.

The Palestinians are also putting in place measures to adapt. The plans and the efforts by the Palestinians to raise financing for a desalination plant in Gaza is one example of an adaptive measure, although also here motivated primarily by the already very troublesome situation in Gaza as it relates to water availabil-ity and quality (Attili 2012). The efforts to establish a plant in Gaza have also been agreed upon as a worthwhile project by the Union for the Mediterranean (UFM), to which both Palestine, Israel, and Jordan (among others) have agreed.[7] Furthermore, many development partners are keen to finance climate change adaptation projects and programmes. UNDP has been leading the efforts in Palestine in relation to the EQA and identified a range of adaptation measures. Messerschmid (2012) argues that this is encouraging government agencies (as well as civil society) to seek a climate change angle on their projects in order to attract more financing for projects.

In Jordan there is also an increasing focus from development partners for climate change adaptation measures. In addition to this, the planned projects in Jordan for desalination in Aqaba[8] are intended to increase supply, as is the Disi project where groundwater is being pumped from a transboundary aquifer shared by Jordan and Saudi Arabia (Namrouqa 2012a, 2012b). The Red–Dead Sea Conveyance weir, which is a study programme where the government of Israel, Jordan, and the Palestinian Authority agreed to study the prospects of bringing water from the Red Sea to the Dead Sea to diminish the decline of the Dead Sea, to develop hydropower, and also to (primarily to Jordan) desalinate water to be used in the Amman region. The project of studying the prospects of a conveyance weir is facilitated by the World Bank and supported by develop-ment partners such as France and Sweden. The aims of the project are centred on a shared vision focusing on three main pillars, which are: 1) Saving the Dead Sea from environmental degradation; 2) Desalinating water and generating energy at affordable prices; and 3) Building a symbol of peace and cooperation in the Middle East.[9] A feasibility study has been developed for the project but currently the process seems stalled for political reasons. According to the Jordan Water Strategy (2008), the country aims to almost double its water budget from around 800 MCM to over 1,600 MCM in 2022. The two key features in this increased supply are the Red–Dead Sea Conveyance and the Disi project. While the full scale Red–Dead Sea is unlikely to happen, since the feasibility studies showed that it is economically too costly, an agreement was reached in December 2013 between Israel, Palestine, and Jordan.[10] This is by some portrayed as a first step of the Red–Dead project, but is in essence a 'water swap' between Israel and Jordan, with the possibility for Palestine to buy additional water from Israel at a price yet to be agreed upon. Jordan will desalinate water close to Aqaba at the Red Sea shore, and then transfer part of that water to Israel (at market price) in the south. In return Israel would sell water (or similar amount) to Jordan from the Tiberias Lake up north at market price. The brine from the desalination plant in Aqaba is intended to be put into the Dead Sea. In addition, Israel agrees to sell water to Palestine in the West Bank. However, much remains to be further developed in relation to the financing of the plant, and primarily the pipe to take the water to the Dead Sea, but also in relation to the price of water for the trade between Israel and Palestine (personal communication, Shaddad Attili 5 February 2014). All of the above-mentioned projects are primarily motivated from Jordan's perspective on its strategy to address its increasing water scarcity, but can also be viewed as adaptation to climate change and increased climate variability. In addition, UNDP is working with a climate change adaptation project focusing partly on its water resources but also with a link to the sustainment of Jordan's progress towards achieving the Millennium Development Goals (MDGs).[11]

It is noted that both significant technical and financial resources are being allocated towards climate change adaptation – although it may not be always be the primary cause of the investments, as is the case with the desalination plants in Israel. The desalination plant being planned for Gaza or the Red–Dead Sea Conveyance Weir (or the parts of it that was agreed in December 2013) is of prime importance in particular for Jordan. None of these investments, which will nevertheless help them to adapt to climate change, had invoked climate change as the prime *motif* for the investments. Rather, the investments address current water scarcity challenges for all the countries (Feitelson, personal communication 2012).

On the organisational level it seems a degree of capacity building of staff working in the environment and water fields is taking place and is primarily supported by development partners. In that sense, there is a certain degree of *adaptation* measures being promoted. However, as alluded to by Messerschmid (2012), it may be questioned if this is a development that is happening because there is funding available for climate change projects (including capacity building) or if it would have happened nevertheless.

Resist

At a national level, a change towards establishing units/departments or similar to address the challenges of climate change is not visible. This may not be a direct resistance, but rather an indirect form of not highlighting the issue from an institutional perspective.

In the societies there seems to be an understanding that climate change is happening and it will affect their water situation (AFED 2009). However, climate change is not particularly high on the political agenda in any of the countries. Sowers et al., in their overview of the MENA region's efforts to promote climate change adaptation, concluded that '. . . a primary, however, remains convincing political leaderships of the urgent need to prioritize adaptation measures' (Sowers et al. 2011: 622). This is not resistance as such towards the arguments being put forth in the global climate change debate, but is a signal of its relative political and societal importance.

Legal agreements

On the legal side it has been noted that in the agreements that currently guide the shared waters in the basin there is scarce provision for dealing with climate variability and even more so with climate change (which would mean increased climate variability). Neither the Palestinian–Israeli interim agreement nor the Israeli–Jordanian agreements include clear provisions for how to address droughts (Jägerskog 2003; Falkenmark & Jägerskog 2010). The mechanisms that are commonly referred to as being the tools to address variability are the JWCs, which exist in both cases. However, when there is a drought in the region, a political conflict has often followed, showing the weakness of the approach. The lack of flexibility in the agreements is clear. In that sense, there has been a *resistance* to acknowledge the frameworks and agreements developed to address transboundary waters. The reasons for this 'resistance' can be many. Often issues are left out of an agreement either because negotiators were not able to identify them at the time of negotiation or because they may for political reasons choose not to address them (Jägerskog 2003). In this case it is unlikely that variability as a factor was something the negotiators were not able to identify since this has always been a constant feature in the Jordan basin. Jägerskog (2003) concludes that for political reasons variability was pushed to the JWC between Israel and Jordan rather than inserting specific text on the issues. That 'constructive ambiguity' also allowed for Israel and Jordan, respectively, to interpret the agreement slightly differently, which is something that has brought challenges since 1994. However, in the Jordan Water Strategy (2008) there seems to be an acknowledgement of the fact that there is a need to address challenges at a regional level, and it points out that Jordan's 'Bilateral and multi-lateral co-operation with neighbouring states shall be pursued through a Regional Water Charter' (Jordan Water Strategy, Chapter 3, Article 36). Whether that view is being reciprocated by Israel or the Palestinian Authority is unclear.

Investments

In terms of investments, in infrastructure there has been a considerable effort to establish desalination plants (primarily by Israel, but efforts are currently underway in Jordan and Palestine). These are, however, efforts and initiatives that will increase the ability of the countries to adapt to increased climatic variability in the region (see also above under 'Adapt').

Subvert

In the region, already volatile and plagued by conflict, it seems as though the issue of climate change adds another layer to the conflict. Messerschmid (2012) concludes that, in many respects, the issue of climate change and its relation to water is just a continuation of the present situation where the status quo is favouring Israel, being the dominant party in the region. That Israel is maintaining a dominance in the water sector in relation to Jordan and in particular the Palestinians has been shown many times by researchers as well as international institutions (see for example: Jägerskog 2003; Selby 2003; Zeitoun 2008; World Bank 2009).

The argument made by Israel is that the climate change in the region will make it harder for them to be flexible in the eventual final status negotiations with the Palestinians. In that sense the climate change arguments could be used for purposes that do not relate to adaptation, but rather as subverting the arguments. However, a counterargument is also emerging within the Israeli discourse relating to how the desalination industry is a potential 'game-changer' in the sense that the production of 'new' water that enters the Israeli water budget presents more options in relation to its water diplomacy with the Palestinians. In the framework of discussions under the lead of the United States, the perspective that Israel may be more forthcoming could gain ground in the larger scheme of negotiations. It is significant that the desalination will add around 750 MCM to the Israeli water budget on an annual basis (Aviram, personal communication 2012). That level will be reached in 2020, but already at this time over 500 MCM should be produced (Israel Water Authority 2010). These developments are also made easier by the relatively recent findings of natural gas in Israel, which provides important energy for the powering of the desalination plants (Hammarskjöld, personal communication 2012). While these gas discoveries make desalination cheaper, there is not much discussion on how the energy use for the desalination plants – as a contributor to greenhouse gases – contributes to climate change.

Discussions and conclusions

The role of the GCC discourse on TWM in the Jordan basin

In the Jordan basin, the major focus is still on water allocation, and water is part of the conflict between primarily Israel and the Palestinian Authority. At the same time there are plentiful efforts, which are not sparked by trying to adapt to climate change but rather are motivated by water scarcity in the countries already in the present situation. However, in the discourse and debate in the basin, climate change is not a major feature, due to a focus on other political issues and the general water scarcity situation. The debate at the national level(s) is quite informed, but climate change is often being perceived to be less important than other political issues and climate change adaptation fails to catch the attention of politicians (Sowers et al. 2011). As noted by Messerschmid (2012), much of the climate change work is being promoted and supported by international actors and is therefore making it attractive for regional actors to work on this too. However, the issue has yet to 'sink in' among the public, as well as the elite.

The role of basin variables

Desalination plants have during the last decade become much more prevalent along Israel's Mediterranean coast, and the Palestinians are aiming to build a desalination plant for Gaza. Meanwhile, Jordan is also working on desalination in Aqaba and construction facilities in the south of the country to harness water from the Disi aquifer, which is already in operation. The development of (mainly) supply-sided solutions to the water scarcity has added adaptation tools for the countries in the region. In the case of Israel, desalination can be seen to be a 'game-changer' in the sense that since desalination provides an additional up to around 30 per cent of its water budget, it clearly opens up for more compromise with its neighbours (primarily Palestine). It is likely, however, that Israel regards the increased volumes in the water budget as something that will cater for future needs (which are certainly there with increasing population, among other factors) rather than making them more amenable to use water as a bargaining chip in negotiations with Palestine. Having said that, it is of course entirely possible – in a grand deal for a potential final agreement – that water may be used as a means to that end. This is the case not least since water is considered less important – politically – than the other issues at the negating table (Jägerskog 2003).

Responses to GCC in the Jordan basin

At the regional level there is sparse action as well as debate. Sowers et al. (2011) noted that there have been a number of programmes aimed at (partly) addressing adaptation issues at the MENA level, which often means that Israel is not included. In the Jordan basin the only environmental NGO that works in a serious transboundary manner – Friends of the Earth Middle East (FoEME), with offices and co-directors in Israel, Jordan, and Palestine – has been highlighting the challenges that climate change will bring and discussed how it will affect the shared water resources (FoEME 2007). FoEME are suggesting sensible proposals, including the need for more flexible agreements to cater for increased variability in flow, something that looks unlikely at present. At the level of the Arab States, UNESCWA is developing knowledge and scenarios to improve understanding as well as preparedness among its member states. As noted above, most of the approaches to the challenges in the region are largely supply-sided and involve the building of desalination plants for the production of 'new' water, but improved efficiency is also a focus in all the countries. It is deemed that this is not primarily driven by climate change concerns, but is as much (if not more) designed to deal with the water scarcity situation already present today. Most of the responses are deemed to be adaptive; there are signs also of resistance, as well as subversion. The lack of acknowledgement and the subsequent lack of proper addressing of climate change concerns in agreements is a sign of resistance in the basin.

Notes

1 FAO AQUASTAT. Online database (2007). http://www.fao.org/nr/water/aquastat/data/query/index.html
2 Palestinian Water Authority. Water Database. Palestinian National Authority, Palestinian Central Bureau of Statistics, 2007, http://www.pcbs.gov.ps/Portals/_pcbs/Water/c6f0eba6–3f6a-41b3-baad-c6b13a24ddc8.htm
3 Informal discussions with people involved in the 'Kerry talks'.
4 Executive Action Team: http://exact-me.org
5 See: http://www.glowa-jordan-river.de/OurProducts/HomePage
6 See: http://www.escwa.un.org/RICCAR/ri.asp?ReferenceNum=RI
7 For info see: http://ufmsecretariat.org/wp-content/uploads/2011/07/Gaza-Desalination-Project-Fact-Sheet-14-May-2012.pdf
8 http://www.emwis.net/thematicdirs/news/2008/08/jordan-red-sea-desalination-plant-be-and-running
9 For more information see: http://web.worldbank.org/WBSITE/EXTERNAL/COUNTRIES/MENAEXT/EXTREDSEADEADSEA/0,,contentMDK:21827416~pagePK:64168427~piPK:64168435~theSitePK:5174617,00.html
10 See: http://www.worldbank.org/en/news/press-release/2013/12/09/senior-israel-jordanian-palestinian-representatives-water-sharing-agreement and also news reporting from: http://www.nytimes.com/2013/12/10/world/middleeast/israel-jordan-and-palestinians-sign-water-project-deal.html?_r=0
11 See: http://www.undp-jordan.org/uploads/projects/pd_1280849702.pdf

References

Allan, Tony, 2001, *The Middle East Water Question: Hydropolitics and the Global Economy*. London and New York: I. B. Tauris.

Alpert, P., Halfon. N., Levin, Z., 2008, 'Does air pollution really suppress precipitation in Israel?' *Journal of Applied Meteorology and Climatology*, 47, 933–943.

Arab Forum for Environment and Development (AFED) 2009, *Arab Public Opinion and Climate Change, 2009*. Arab Forum for Environment and Development.

Assaf, Karen, al Khatib, Nader, Shuval, Hillel, and Kally, Elisha, 1993, *A Proposal for the Development of a Regional Water Master Plan*. Center for Research and Information (IPCRI), Jerusalem, Israeli/Palestine.

Beck, Martin, and Collet, Lea, 2011, *Jordan and Climate Change – Effects, Perceptions and Adjustment Measures in Jordan*. Konrad Adenauer Stiftung, see: www.kas.de/jordanien

Cooley, H., and Gleick, P.H., 2011, 'Climate-proofing transboundary water agreements', *Hydrological Sciences Journal*, 56(84): 711–718.

De Stefano, L., Duncan, J., Dinar, S., Stahl, K., Strzepek, K., and Wolf, A., 2010, *Mapping the Resilience of International River Basins to Future Climate Change-Induced Water Variability*, Water Sector Board Discussion Paper Series, World Bank, Paper no. 15.

El-Khoury, Rami, 2012, *Desalination and the Arab-Israeli Issue*, Al-Jazeera Blogpost: http://www.aljazeera.com/indepth/opinion/2012/08/201281373146974754.html

Falkenmark, M., and Jägerskog, A., 2010, 'Sustainability of transnational water agreements in the face of socio-economic and environmental change', in Earle, A., Jägerskog, A. and Öjendal, J. (eds.), *Transboundary Water Management: Principles and Practice.* London: Earthscan.

Feitelson, Eran, and Rosenthal, Gad, 2011, *Desalination, Space and Power: The Ramifications of Israel's Changing Water Geography.* GeoForum (in print).

Feitelson, Eran, Tamimi, Abdelrahman, and Rosenthal, Gad, 2012, 'Climate change and security in the Israeli-Palestinian context', in *Journal of Peace Research,* 49(1): 241–257.

Fishhendler, I., 2004 'Legal and institutional adaptation to climate uncertainty: a study of international rivers', in *Water Policy,* 6(4): 281–302.

FoEME, 2007, 'Climate change: a new threat to Middle East security'. Prepared for the United Nations Climate Change Conference Bali, Indonesia EcoPeace/Friends of the Earth Middle East Amman, Bethlehem, and Tel-Aviv, December 2007.

FoEME, 2012, 'Take me over the Jordan: concept document to rehabilitate, promote prosperity, and help bring peace to the Lower Jordan River Valley', EcoPeace/Friends of the Earth Middle East, Amman, Bethlehem, Tel Aviv.

Haddadin, Munther, 2000, 'Water issues in the Hashemite Jordan', *Arab Studies Quarterly,* 22(2), 63–77.

Haddadin, Munther, 2001, *Diplomacy on the Jordan: International Conflict and Negotiated Solution.* London: Kluwer Academic.

IPCC, 2007, 'Climate change 2007: the physical science basis', in Solomon, S., Qin, D., Manning, M., Chen, Z., Marquis, M., Averyt, K. B., Tignor, M., and Miller, H. L. (eds.), *Contribution of Working Group I to the Fourth Assessment Report of the Intergovernmental Panel on Climate Change.* Cambridge, UK and New York: Cambridge University Press.

Israeli Water Authority (IWA), 2010, 'Long-term national master plan for the water sector: policy document', draft, December, Tel Aviv (Hebrew).

Jägerskog, A., 2003, 'Why states cooperate over shared water: the water negotiations in the Jordan River basin', PhD dissertation, Linköping Studies in Arts and Science: Linköping University.

Jägerskog, A., 2008, 'Functional water co-operation in the Jordan River basin: spillover or spillback for political security', in Brauch, Hans Günter, Oswald Spring, Ursula, Grin, John, Mesjasz, Czeslaw, Kameri-Mbote, Patricia, Chadha Behera, Navnita, Chourou, Béchir, and Krummenacher, Heinz (eds.), *Facing Global Environmental Change: Environmental, Human, Energy, Food, Health and Water Security Concepts.* Hexagon Series on Human and Environmental Security and Peace (Vol. 4). Berlin: Springer-Verlag.

Kolars, John, 2000, 'The spatial attributes of water negotiation: the need for a river ethic and advocacy in the Middle East', in Amery, Hussein A. and Aaron Wolf (eds.), *Water in the Middle East: A Geography of Peace.* Austin: University of Texas Press.

Messerschmid, Clemens, 2012, 'Nothing new in the Middle East – Reality and discourses of climate change in the Israeli-Palestinian conflict', in Scheffran, Jürgen, Brzoska, Michael, Brauch, Hans Günter, Link, Peter Michael, and Schilling, Janpeter (eds.), *Climate Change, Human Security and Violent Conflict – Challenges for Societal Stability.* Hexagon Series on Human and Environmental Security and Peace (Vol. 8), Berlin: Springer-Verlag.

Namrouqa, Hana, 2012a, 'Construction work on Disi project nears completion', *Jordan Times,* accessed 27 November 2012 at: http://jordantimes.com/construction-work-on-disi-project-nears-completion

Namrouqa, Hana, 2012b, 'Implementation of Jordan Red Sea Project to begin early next year', *Jordan Times,* 27 March 2012, accessed 20 March 2014 at: http://mideastenvironment.apps01.yorku.ca/2012/03/

Namrouqa, Hana, 13 May 2013, 'Climate change may reduce water availability, damage agriculture in Jordan', *Jordan Times*, 13 May 2013, accessed 16 February 2015 at http://jordan times.com/climate-change-may-reduce-water-availability-damage-agriculture-in-jordan

Oamri, Raed, 2011, 'Too early to declare water emergency', 24 February 2011, *Jordan Times*, accessed 1 December 2010 at http://jordantimes.com/index.php?news=32238

Phillips, D.J.H., Attili, S., McCaffrey, S., and Murray, J. S., 2007, 'The Jordan River basin: potential future allocations to the co-riparians', *Water International, 32*: 39–62.

Rinat, Zafir, 17 May 2013, 'The plan is intended to ecologically rehabilitate the neglected river', *Haaretz*.

Selby, J., 2003, 'Dressing up domination as "co-operation": the case of Israeli-Palestinian water relations', *Review of International Studies, 29*(1): 121–138.

Selby, J., 2013, 'Cooperation, domination and colonisation: the Israeli-Palestinian Joint Water Committee', *Water Alternatives, 6*(1): 1–24.

Shapland, Greg, 1997, *Rivers of Discord: International Water Disputes in the Middle East.* London: Hurst & Co.

Sowers, Jeannie, Vengosh, Avner, and Weinthal, Erika, 2011, 'Climate change, water resources and the politics of adaptation in the Middle East and North Africa' in *Climate Change,* 104:599–627.

Trondalen, Jon Martin, 2009, 'Climate changes water security and possible remedies for the Middle East', Paris: UNESCO, accessed 16 February 2015 at http://unesdoc.unesco.org/images/0018/001818/181886E.pdf

UNDP PAPP, 2010, *Climate Change Adaptation Strategy and Programme of Action for the Palestinian Authority.* Jerusalem: United Nations Development Programme, Programme of Assistance to the Palestinian People.

UNEP, 2012, *Vulnerability Assessment of Freshwater Resources to Climate Change: Implications for Shared Water Resources in the West Asia Region.* UNEP: Nairobi.

'Water for life: Jordan's water strategy 2008–2022', accessed at http://www.idrc.ca/uploads/user-S/12431464431JO_Water-Strategy09.pdf

Wolf, Aaron T., and Hamner, Jesse H., 2000, 'Trends in transboundary water disputes and dispute resolution', *Water for Peace in the Middle East and Southern Africa.* Geneva: Green Cross International.

World Bank, 2009, 'West Bank and Gaza: Assessment of restrictions on Palestinian water sector development', Sector Note, Middle East and North Africa Region – Sustainable Development, Report No. 47657-GZ. Washington, DC: The International Bank for Reconstruction & Development/World Bank.

Zeitoun, M., 2008, *Power and Water in the Middle East: The Hidden Politics of the Palestinian-Israeli Water Conflict.* IB Tauris, London.

Zeitoun, M., and Jägerskog, A., 2009, 'Confronting power: strategies to support less powerful states' in Jägerskog, A, and Zeitoun, M., (eds.), *Getting Transboundary Water Right: Theory and Practice for Effective Cooperation.* Report Nr. 25. SIWI, Stockholm.

Zeitoun, M. and Mirumachi, N., 2008, 'Transboundary water interaction I: reconsidering conflict and cooperation', *International Environmental Agreements, 8*: 297–316.

Other Sources

Israel Water Authority, 2010, Presentation made on October 4, 2010, at the EuroMed Conference in Tel Aviv, by Abraham Tenne, Head of Desalination and Water Technologies at Israel Water Authority and Chairman of the Water Desalination Administration.

Personal Communication

Attili, Shaddad, personal communication, Minister, Palestinian Water Authority, Ramallah, 11 January 2012 and 5 February 2014, Amman.

Aviram, personal communication, Ambassador and former Israeli Head of Multilateral, Jerusalem, 17 December 2012.

Bromberg, Gidon, personal communication, Israeli Director of Friends of the Earth Middle East, Jerusalem, 17 December 2012.

Carmi, personal communication, Advisor, Neogiation Support Unit, Water File, PLO, Jerusalem, 12 January 2012.

Feitelson, Eran, Professor, Hebrew university of Jerusalem, personal communication, Jerusalem, 11 January 2012.

Hammarskjöld, Ellinor, personal communication, Ambassador of Sweden to Israel, Tel Aviv, 10 January 2012.

Mimi, Ziad, personal communication, Professor, Bethlehem University, Bethlehem, 12 2012.

4   The Mekong River basin

Introduction

Southeast Asia in general, and the lower Mekong basin in particular, has been pointed out as one of the areas in the world that is most prone to extreme changes in weather conditions as a result of global climate change, while at the same time harbouring agricultural and natural resource-based livelihoods that makes it excessively vulnerable to such changes (Cruz et al. 2007; cf. Kuroda & Stern 2009).[1] In addition, poverty is widespread and infrastructure preparedness – physical or institutional – for extreme events are limited in most countries, further enhancing vulnerability.[2] Finally, as we argued in the introductory chapter, GCC-consequences are 'water-driven', and this region is to a large extent defined by the water regime in the Mekong River, its wetlands, and its tributaries providing *inter alia* water, fertile soil, transport routes, irrigation, and ample fisheries throughout the lower basin. There seems to be massive justification for taking GCC-induced alterations to the climate seriously in the region as various studies – although differing in scope and timing is differing – predicts future critical GCC-impact in the region (Cruz 2007; Eastham et al. 2008; Johnston et al. 2010; TKK & SEA-START RC 2009). These predictions are at large further predicted by meta-climate studies using other methodologies (Lacombe et al. 2013).

The Lancang-Mekong River is situated in southwestern China and mainland Southeast Asia. It is one of the major rivers in the world, ranking number eight in terms of flow, covering an international basin of six countries[3] and 795,000 km^2 that are inhabited by some 80 million people. One defining feature of the Mekong water regime is that the seasonal variations in flow are extreme, where the rainy season flow is up to 25 times as high as the dry season one. It flows up in the Himalayas, China, touches upon Myanmar, before it reaches the lower Mekong basin, where Laos and Thailand share the river before it reaches the plains of Cambodia, and finally the delta in Vietnam. Originating in the Himalayas, through China the river is fast flowing with steep mountainsides and limited agricultural potential. Instead the hydropower potential is huge as it is ascending some 4,000 metres, often in a dramatic terrain. In Laos, it meanders through mountains, and the ample rainfall in the mountainous areas to the east of the mainstream generates an entire landscape of tributaries with high hydropower potential, as well as creating a base for upland minorities and their traditional livelihoods. In Cambodia, the landscape flattens and the river often floods the plains, and at times the river flows 'upwards' into the various tributaries and ample wetlands. Especially crucial is when the Tonle Sap Lake fills up, reaching a size five times larger than its dry season extension. In the delta in Vietnam, finally, the river crisscrosses in a vast system of natural and man-made canals, at places just a few metres above sea level. Here it gives rise to both highly efficient rice production and fisheries, although the sectors' priorities occasionally compete and both being highly vulnerable to disturbances (Öjendal 2000). Hence, the importance of the Mekong basin is neither only the river nor the water in itself, but the ecosystem and the resource base that it sustains. Some 80 per cent of the basin's inhabitants fully or partly survive through livelihood-based natural resources utilisation made possible by the diverse and dynamic ecosystem the water regime keeps up (Fox & Sneddon 2005; MRC 2010c).

The Greater Mekong basin covers one of the poorest regions in the world. Although harbouring major economic inequalities, the rural areas – which in the basin often coincide with the stretch where the Mekong mainstream is situated – have paradoxically the highest instances of poverty and the most vulnerable situation in terms of water access. In China, Thailand, and Vietnam, the areas adjacent to the river are marginal in relation to the respective national capitols and regional urban centres. In Laos and Cambodia, the river basin covers a major part of the countries, but both are entrenched low-income countries.

In the face of hastened pressure for modernisation (and, indeed, increased success in pushing growth and modernisation), increased water withdrawal for dam building, as well as for intensified agriculture, and to a booming commercial plantation industry is evident across the basin (Johnston et al. 2010:8–9; 14–15). Rivalry for water is deepening, and incompatible interests in water (and indeed the threat to the sustainability of the overall ecological system) is spreading among countries and between sectors. While the resource base is strained, the demand for economic development and poverty reduction is urgent, producing a dilemma for policymakers. The Mekong River basin and its seemingly abundant water resources has become a tense focus area for economic growth.

At the lower basin level, there is a long-established River Basin Organisation (RBO) in the Mekong River Commission (MRC, see further below). In 1957, the institutionalised basin-wide cooperation was initiated through the Mekong Committee, and was then disrupted for decades by war and violence, but reinvigorated 1995 in a potentially potent framework agreement named the MRC-Agreement, which is mandated to promote the cooperation of all riparians in:

> fields of sustainable development, utilisation, management and conserva-
> tion of the water and related resources of the Mekong River Basin includ-
> ing, but not limited to irrigation, hydro-power, navigation, flood control,
> fisheries, timber floating, recreation and tourism, in a manner to optimise

the multiple-use and mutual benefits of all riparians and to minimise the harmful effects that might result from natural occurrences and man-made activities.

(MRC 1995: Art 1)

More specifically, it is also tasked to specifically address joint environmental issues since it shall work to:

protect the environment, natural resources, aquatic life and conditions, and ecological balance of the Mekong River Basin from pollution or other harmful effects resulting from any development plans and uses of water and related resources in the Basin.

(MRC 1995: Art 3)

There is institutionalised preparedness to deal with the governance challenges emerging in the process, as the first imperative for proper basin governance is therewith in place. Or as de Stefano et al. state:

the make-up and design of treaties, RBOs, and related provisions are anticipated to be particularly important in assuaging potential interstate conflict or country grievances, which may be caused by an increase in inter-annual water variability due to climate change.

(De Stefano et al. 2012: 197)

The major drawback would be that China is not a signatory to the Agreement; neither is Myanmar. In spite of the absence of China and Myanmar, the Mekong basin and its riparians possibly harbour a competitive advantage due to its long-established institutionalisation of water governance at basin level, and, as we shall see below, already having initiated several projects on adapting to GCC. While GCC-projects are tentatively established at basin-level, we argue, more hands-on governance are primarily carried out at national level, more or less institutionalised in the various ministerial systems, although also these processes are at times vague.

As a further counter-image to any simple perceptions on cooperation on basin-level, it must be noted that political relations among the riparians are historically complex and recently embittered in several ways. Although the MRC provides a major institution for river basin management (Öjendal et al. 2012), it has mostly been successful in technical matters and only to a small extent crafted political consensus around basin management (Hirsch & Mœrck-Jensen 2006; Lazarus & Dore 2009). Moreover, there is an intensive modernisation pressure in the region, creating a rivalry for the still available resources and possibly also inflated plans made for strategic reasons to 'safeguard' sufficient water for emerging national plans. To a major extent, this process is eyeing the Mekong water resources due to its huge hydropower potential feeding the energy-hungry riparian economies (and to a minor extent for large-scale irrigation and/or sustaining cash-crop plantation ventures). Consequently, there are major hydropower schemes being planned and gradually emerging (Hirsch 2010; Johnston et al. 2010), severely straining the smooth long-term planning and easy access to a smooth and balanced cooperation around Mekong development. Hence, there is urgency around water management which provides an impediment to prioritise incremental and future-oriented issues such as global climate change; the long-term planning risks to be crowded out by the promises of short-term economic gains and immediate political issues, or even going a step further, the GCC discourse may be crafted as to fit various interests and plans.

As we shall see below, all riparians are physically affected by the GCC and its impact on people and livelihoods, and there are major interests involved in the anticipated change. In more detail, Vietnam has a major stake in protecting the low-lying delta area from salt water intrusion; this is already a major threat to the agricultural productivity of the area, but with rising sea level and more extreme climate, the risk is considerably enhanced. *Cambodia* protects its Tonle Sap system – upheld by annual floodings – which sustains a large share of the countries' water regime, and underpins its rich fisheries. *Lao PDR* aims for major hydropower development and tries to not get GCC 'in the way' for those major infrastructure development, whereas *Thailand* is battling with both regional poverty in the Korat Plateau and to get water management right in its share of the basin under intensive public pressure originating from GCC-induced alterations, and other sources of contention.

The issue at stake here – how regional water governance responds to GCC – emerges as a process that is multidimensional and paradoxical, and where questions are nested and elusive, drawing on underlying interests trying to 'define' the issue at stake and how do deal with it. In this volume, as well as in this chapter, we apply an analytical framework that sorts reactions to GCC into the processes of adaptation, resistance, and/or subversion. Obviously these can (and do) take place simultaneously at different levels and played out by different actors. In the Mekong there is, however, a river basin authority (MRC) with a rather distinct actor capacity, whose reactions can be traced. Beneath this regional level, other actors pursue (entirely in line with the overall theoretical understanding of the volume) a plethora of strategies for how to meet the rather massive GCC discourse. Hence, below, we will review the projections on the GCC-induced alterations, trace the discourse on GCC in the Mekong basin, describe the river basin management arrangements, and analyse the current efforts at handling the emerging issue(s) at regional and national level. Finally, we conclude on to what extent GCC is *adapted to, resisted, and/or subverted* in, and through, the lower Mekong basin governance modalities.

Global climate change impact – projections and consequences

The weather pattern in the region is driven by two different monsoon systems that are interacting in a complex way (MRC 2010c). These monsoons are powerful and balance each other delicately. As such it is unusually sensitive to changes in global weather patterns:

Although climate change is a long-term process of incremental change in regular climate patterns, the LMB is projected to experience increased magnitude and frequency of extreme events such as floods and drought which can occur at any time.

(GCC-USAID 2013: 24)

Similar projections were issued already at the turn of the century (IPCC 2001), and this discourse has since been effectively grounded in the region. For instance, ADB has stated that some 'famous' floods and droughts 1997, 1998, and 2000 are examples of the increased unpredictability brought by the climate change in the region (ADB 2009, taken from Johnston et al. 2010). WWF (2009) claims that GCC is already 'being felt' in the region. Similar sentiments were repeated in virtually every interview made for the writing of this chapter – from officials and civil society representatives alike – and commonly argued in policy circles as justification for moving the frontiers on the preparation for change in the basin. There is no doubt that the discourse on GCC has made a wide and deep inroad into the debate on Mekong basin management (cf. Thapan 2009; cf. Gerlak & Schmeier 2014).

Turning to material 'facts' (or at least to what are presented as such), the weather variability in the Mekong basin is extreme also under 'normal' circumstances, with massive rainfall but with major seasonal and local variations. Highest rainfall occurs in the western parts of the region in Lao PDR and Vietnam, reaching (or exceeding) 2, 500 mm/year, while large tracts of northeast Thailand and of the Delta see less than 1,200 mm/yr. (GCC-USAID; MRC 2010c). These rainfall figures need, however, to be understood in their context since areas with 2,000 mm/year usually seems healthy, but these areas may nevertheless experience both flooding and drought, even within the same year. As a result of uneven rainfall during the year, the flow in the rainy season is up to 25 times as high as during the dry season, creating a very special water regime, with unique livelihood systems (cf. Sneddon & Fox 2005).

By 2030, projections are:[4] rise in mean temperature by 0.8 degrees; 13.5 per cent rise in wet season precipitation; increased severity, duration and frequency of flooding; and, disturbances/reduction in agricultural output and capture fisheries (Eastham et al. 2008; MRC 2009a). In addition, it is commonly considered a scientific fact – as stated in a USAID report – that variations will be dramatic: 'In the Lower Mekong Basin, temperature increases are expected to reach an average 3–5°C by the end of the century, however some pockets of the basin are predicted to experience much larger increases' (2013: 78). Johnston et al. (2010: 16) are, with a different methodology, mentioning a rise in 2–3 degrees by the turn of the century, a considerably rising demand for irrigation (as evaporation increases with rising temperatures), and possible temporal and spatial change in rainfall (although a rather modest average change). Projections vary, but the discourse on fundamental change is solidly established. As a result, disaster preparedness is high on the list of adaptation activities in many action plans regionally (see below).

The changes are not only predicted to be unusually harsh in a biophysical way; the lower basin alone supports some 65 million people, of which a large portion is poor and/or dependent on the existing water regime. Hence, the altering conditions are also expected to have severe implications for the various and diverse livelihood systems.

> Natural resources are essential to rural livelihoods – dependence likely to increase for the poor in the coming decades. This reliance reflects the acute sensitivity of rural households to adverse weather events, such as floods and droughts, as well as to degradation of the natural environment.
>
> (GCC-USAID 2013: 18)

In addition, the combination of predominantly traditional livelihoods on the one hand, and the increasing degree (or increasing mix as it were) of modernised and commercial farming on the other, underscores the vulnerability:

> Subsistence based systems are inherently integrated with natural systems and benefit from their diversity and resilience to climate related shocks. However, natural systems are being degraded in the LMB, partly due to the shift from subsistence to commercial agriculture. As farming systems move along this continuum they are becoming less diverse, more intensive, and less resilient to climate change.
>
> (GCC-USAID 2013: 18)

In essence, this feeds into an older debate on change and sustainability, which is getting increasingly unavoidable (Öjendal 2000; Fox & Sneddon 2005). Natural science tells us that GCC is now tangible in the basin and that it will be distinctly felt already in the medium-term future, and be severe in the long term. Moreover, locally and on bad years, it will create disastrous effects, especially placing a severe burden on traditional livelihoods and poor segments of the societies in question. While this is not necessarily a scientific 'fact', it is what emerges from the most knowledgeable studies for this region the last decade. Let us view how this is debated in the region.

The emerging discourse on global climate change in the Mekong basin

In addition to the scientific 'facts' above, GCC is also tangible in the region as a more-public discourse, here illuminated through opinions voiced at/through the leading newspapers in the region.[5]

While there are many interests and actors who are picking up on GCC, the issue is neither (yet) deeply politicised, nor widely spread among the public. ('Want to talk about what happens in 30 years? Come back and talk to my son', as one report cited a poor peasant's response when asked about GCC). Nevertheless, there is now a sense of urgency in the region and a growing discussion on what to do about it and who to blame for possible inadequate preparations.

The people of Thailand shiver under an unexpected cold spell, the Indo-
nesians rescue survivors from a bout of storms, and Filipinos prepare with
dread for another year of worsening typhoons, the Burmese are trying to
restore their lives and livelihoods after the devastation of Typhoon Nargis,
and at the lower end of the Mekong River, Cambodians are fighting a rising
number of Malaria cases.

(Shailendra Yashwant, *The Nation*,
23 January 2009)

Kuroda and Stern – from the vantage point of the 'Stern report' and others –
note in an article headlined 'Why Global Warming Could Make or Break South-
East Asia', that a change of attitude is necessary for 'saving' the region:

Few regions on Earth throw this tension into sharper contrast than
south-east Asia, where many nations are highly vulnerable to the effects
of global warming while also having the chance to develop low-carbon
economies. . . . By the end of this century temperatures in south-east Asia
will rise significantly, tens of millions will experience water shortages, rice
production will decline, and large swaths of forests will disappear. Rising
sea levels will force the relocation of millions of island dwellers and coastal
communities, and there will be a surge in dengue, malaria and other diseases.

(Kuroda & Stern 2009)

Also 'modest' voices emphasise the special urgency of GCC in the region:
'The Mekong region is globally quite unique' (Ketelsen in *Bangkok Post*, 9 April
2013). And, 'If you look at the IPCC-reports this area is a black hole. All the
parameters are more extreme (. . .) temperature, rainfall and flooding' (Carwe-
Reid in *Bangkok Post*, 9 April 2013). From a more political angle, Abhisit Vejja-
jiva, the former Prime Minster of Thailand, said, 'We have no time to waste. Let
us work together to combat climate change and save this planet for our children
[and] grandchildren' (*The Nation*, 30 August 2010), whereas Suwit Khunkitti,
then-minister of Natural Resources in Thailand, stated that climate change is
to blame for the Mekong River 'drying up' (*Asia News Network*, 9 April 2010).
   One strand of concern is the pondering of 'hot-spots',[6] either as a way of preparing for local change, or as a way of causing a sense of urgency, claiming that
most of these hot-spots 'will experience significant decreases in yield' (*Bangkok
Post*, 9 April 2013). Critical sectors are also noticed, such as fisheries (*The Nation*,
28 October 2010), food production and transportation (*South China Morning
Post*, 30 September 2009), as well as pointing out the obvious local aspect of
this dilemma, calling for '. . . localizing climate change research' (*South China
Morning Post*, 6 June 2008).
   As is pointed out elsewhere in this article, development debates – especially
the dam-building issue and the management aspects of these process – are inevitable drawn in, where various arguments and their underlying vested interests
are borrowing legitimacy and strength from the GCC discourse, 'proving' their
points. For instance, the recent flooding in Thailand is occasionally blamed

on 'human failures', arguing leadership incompetence (Pakamas Thinpanga, *The Nation*, December 2012). While dams occasionally are argued to 'solve' the increasing erratic weather pattern (by reducing impact of high rainfall, and mitigating spells of drought), Pianporn Deetes said that 'There are better ways to meet water and energy needs and the climate change challenge' (*The Nation*, 2 April 2010), also reminding us that dams do not work in such a smooth way as to curb highs and lows, but rather in their own technical and economic rationale often serve to acerbate the original problems.

Two other topics dominate the general discourse. The first are the frequent calls for enhanced regional cooperation. This argument was, for instance, stated by Soriano in *The Nation* in no unclear words: 'Climate change demands that Asean's future is cooperative', and he goes on to argue for support that Asean and its 'Asean Working Group on Climate Change' (AWGCC) take a firm regional stand for low-carbon economies to emerge, noting that the rural poor all over Southeast Asia, as well as a number of their megacities, are among the most vulnerable to accelerating GCC (1 May 2013). It is commonly recognised that this is neither a national problem nor one with national solutions; 'Obviously, climate change is an issue that is bigger than these organisational or political concerns. It threatens the common future of Asean's people' (*The Nation*, 1 May 2013). Sunchindah notes further that also within, Asean policymakers claim that '. . . stronger inter-governmental policy coordination is urgently needed . . .' (*The Nation*, October 2009). And although these calls for deepened coopera-tion are referring to Asean summits and working groups, they are not issued by those with policy-making powers. Rather, this radical part of the discourse is on the outside talking in, with few references to the actual institutions aimed to deal with this, and even less reflecting on which interests it challenges. As there are no regional action plans to follow, arguments often falls to the ground without being picked up in a broader debate. In the absence of strong regional initiatives, some raise the pressure, branding GCC '. . .the greatest environmental threat of all time . . .' (Shailendra Yashawant, *The Nation*, October 2009).

The other recurring topic is the severe knowledge deficit along several dimen-sions. Available knowledge is elusive and sometimes speculative (Johnston 2010), and it is mainly held by professionals, limiting the power of the discourse, prov-ing that knowledge in itself does not produce a discourse, and even less produc-ing operational strategies. 'The problem is', *The Christian Science Monitor* stated, 'nobody knows the impact of climate change' (6 June 2008). Another issue of the same newspaper asked rhetorically, 'How to fight a rising sea?' (November 2007). The relative ignorance on impact and the lack of a distinct plan of action is of course a truism in the field of climate change, where uncertainty is the name of the game (cf. Shackley & Wynne 1996).

It is striking that the entire discourse – glimpses of which are reviewed above – is carried by professionals, outsiders, and/or lower level policymakers; neither the executive political sphere nor the public have engaged in a major way. As such, the general discourse remains 'light' with only minor impact and few political controversies. While this is the state of the art, it is bound to change rather soon. Given the dire prediction from the natural sciences reviewed above, there seem to be few options to trying to govern the changes through political decisions and institutional strengthening, which the three following sub-sections will address.

Institutional arrangements in the Mekong governance

Mekong River planning and management has its roots in the immediate post-WWII era and has gone through a range of phases and political negotiations. From initially having been a flood management control scheme, it grew during the 1960s into plans for major hydropower schemes (MS 1970; cf. Öjendal 2000). This was, subsequently, disrupted by major warfare in the region and civil strife in Cambodia, and it was only in 1995 that regional systematic water governance was again on the agenda with the ambitious 'MRC-Agreement' and its deep commitment to regional dialogue and sustainable use of the basin's resources (MRC 1995; cf. Kranz, Menniken, & Hinkel 2010). Since then, the MRC have been developed, containing a regional political decision-making level, a policy-implementation level, and a large secretariat with high technical capacity and several ambitious programs under its wings. As such, the MRC is expected to be the body where riparians meet for discussing significant issues with management of joint natural resources in the basin, and its secretariat providing administrative services and technically sound basis for decisions.

The lower Mekong riparian countries have received international praise and award for their sincere take on trans-boundary governance, primarily acted out through the MRC. This is often seen as 'successful' and a case to learn from (cf. Jacobs 1995). The most visible results so far of the MRC cooperation are extensive data gathering and dissemination of information concerning the basin's ecological and physical systems, as well as its role as a platform for communication on joint issues. In addition, it has been given praise for an early, systematic, and efficient approach to the GCC (Kranz, Menniken, & Hinkel 2010).

However, this claimed efficiency of the MRC has come increasingly under criticism:

> The Mekong River is a good example of an international river basin that involves multiple sectors and actors and thus needs integrated management. The Mekong River Commission (MRC) has partly adopted this task, but faces many constraints.
>
> (Keskinen et al. 2008: 207)

At large, politicised riparian interests occasionally fail to match with de-politicised regional managerial approaches based on a technical base. Dore and Lazarus make the point that:

> MRC has too often been absent from or silent about substantial decisions being taken on water resources development in the basin. As pointed out

earlier, MRC secretariat has had little involvement and usually very limited information about the hydropower development on the Mekong River mainstream in China, and on tributaries in Laos and Vietnam.

(Dore & Lazarus 2009:16)

There is a lack of commitment by member states, which presents a mismatch with the ambitious MRC-Agreement and limits its efficiency (cf. Hirsch & Mœrck-Jensen 2006). For instance, there is a marked resistance among the riparian states to cede any sovereignty over the shared resources in the basin. This is, however, an institutionally sanctioned position, since the 1995 Mekong Agreement only calls for '. . . cooperation on the basis of sovereign equality and territorial integrity in the utilization of the water resources of the Mekong Basin' (Article 4, quoted in Hang 2008: 4). Having struggled to establish a certain authority in the basin, and partly failed, there is now an agreement that the workload of the MRC will be slimmer and that major tasks will be decentralised to national level. A major question hangs over the future identity of the MRC as a regional water governance agency.

Recently, the controversy surrounding the Xayaburi dam (the first mainstream dam to be constructed in the lower Mekong basin) has fuelled the critical debate on the capacity and centrality of the MRC (cf. Dore & Lazarus 2009). In this case, it has neither been the key institution for technical aspects of planning and assessing, nor has it been (fully) capable of being a mediator in the political debate among the riparians. However, as Öjendal and Mœrck-Jensen point out, the sheer existence of MRC made it impossible to fully prevent the Xayaburi controversy from entering public debate and thereby being subject to political deliberations (2012). In spite of this, the preparation for the next mainstream dam – the Don Sahong in Laos, close to the border to Cambodia – seems to move forward, partly, but not fully, within the framework of the MRC, and certainly with lots of regional tension.

Another controversial issue is the role of China in the Mekong/Lancang cooperation. It was never a member of the institutionalised river basin cooperation in the MRC, although (or perhaps because) it is a regional hegemon in the extreme (large, powerful, and upstream). It has already built several major dams on the upper mainstream (called 'Lancang' in Yunnan/China), and has major plans for future dambuilding (Hirsch 2010; Magee 2012). In spite of China having been an observer and dialogue partner in the MRC, dambuilding plans have been done without consultation with other riparians. However, in the past decade, China has opened considerably vis-à-vis its Mekong neighbours and engaged in more solid information-sharing. China has increasingly been engaged as a dialogue partner with the MRC, and this has led to a limited but steadily growing degree of information-sharing between upstream and downstream countries. A Foreign Ministry spokesperson said in March 2010:

China makes reasonable and sustainable exploitation of the water resources in a responsible way and gives full consideration the reasonable concerns of downstream countries. Over the past years, we have kept close communication and cooperation with GMS countries and have provided them with hydrologic data of upstream water on a regular basis.

(Zhu 2011: 462)

This was put to more explicit use and international recognition at the Hua Hin meeting in 2010, when commitments were officially made to share information on dry season water flows via the MRC.[7]

When GCC is entering the basin and growing in significance, the MRC is the given 'actor' to deal with it. So, the MRC constitutes a classic RBO, with a long history and far-reaching governance ambitions and potential. It does, however, also experience difficulties, as reviewed above. In addition to balancing regional interests and sectoral demands, and as expressed by its former CEO: 'Superimposed on these [challenges] are the future effects of more extreme floods, prolonged drought and sea level rise associated with climate change'. In this, he reiterates the entire rationale for this volume, giving voice to concerns as to whether RBOs, and the agreements they are based on, can take the additional pressure of GCC and its inevitable politicisation (Drieschova et al. 2008). This is at the very least a pertinent question for this basin cooperation.

To sum up, there is long and established institutionalised cooperation over basin development, and there is a regional champion in the MRC, with high technical capacity and modalities for political dialogue, which also is enjoying a basic political legitimacy. There is also a complex governance situation at hand with limited political buy-in, made even more difficult due to massive pressure for modernisation on national level epitomised in the urgent hydropower plans, largely fed by private-sector investments and expectations of high profits. While widely recognised, the GCC is adding further pressure on the governance situation that is already strained. It is far from certain that GCC will be duly respected and addressed to the extent it deserves given the projections and the obvious vulnerabilities. Let us see how the GCC projections and the various GCC initiatives have so far impacted on basin level governance.

Dealing with GCC at basin level in the Mekong

The MRC has initiated an ambitious GCC-project through the 'Climate Change and Adaptation Initiative' (CCAI) (MRC 2009b), commissioned a number of technical reports (Eastham et al. 2008; MRC 2009a; MRC 2010a; MRC 2010b), and is expecting GCC to be a major issue to deal with in the short as well as in the long term (Interview, Senior Manager, MRC). The understanding of an emerging GCC and the necessity of dealing with it is widespread among managers at the MRC, at the national Mekong committees, as well as in the 'professional discourse' on emerging challenges for water governance in the Mekong basin (as above). At policy level, the MRC sees itself being in a favourable position to perform some key tasks of regional adaptation to GCC (MRC, Senior Manager interview),[8] as well as to manage future demands of adaptation.

Most likely, considerable funding will be made available for adaptation to GCC in the region, and one of the obligations of the MRC is to coordinate activities and interventions under this heading (Interview, Senior Manager, MRC). The CCAI flagship project at the MRC was initiated 2008 and is designed to run up to 2025. It contains ambitions at understanding GCC in the basin, build capacity to deal with it, making strategies and plans for how to address adaptation, and create a regional network on the issue (CCAI 2010). Its initial phase is at large financed by AusAid, and endorsed by all member countries. It is under this project that the reports on GCC (2009a; 2010a; 2010b) have been produced by the MRC.

There are, in addition, several other related projects initiated by/through other actors such as 'Mekong Adaptation and Resilience to Climate Change' run by USAID,[9] and the joint financing of 'Study on Climate Change Impact Adaptation and Mitigation in Asian Coastal Mega Cities' by ADB, World Bank, and JBIC. These are technical reports with limited ambition to reflect upon how the governance situation is impacted and how the institutional responses are being designed. As a result: 'Technocratic policy informing and making channels are gaining strength across the Basin, however, transparency and associated issues of "good governance" remain a challenge' (USAID 2013:156). This feeds into a political culture where controversial discussions already are subdued and elusive, and where public debates are avoided (cf. Gerlak and Schmeier 2014).

From this general awareness to actual adaptation may, however, be a far cry. The CCAI has been slow to start, and at the time of interviews, 2012, personnel were missing, progress was unclear, and the expected outcomes in the short to medium term were played down by all parties (Interviews with senior managers, MRC, Vientiane 2012). Several interviewees expressed the sentiment that this was a start-up phase and that we need to feel our way forward. Also, donors admitted that it was difficult to see what would actually come out of it in the short to medium term, and even more difficult to have a distinct strategy for medium to long term (Aid official Vientiane 2012). By January 2014, a Danish review of the progress of the MRC work concluded that the CCAI was one among very few projects that could not reasonably reach its objectives in time.

Overall, MRC has taken a technical/functional approach with a pragmatic political positioning and predictably with a limited impact so far. It is facing two overall dilemmas: *Firstly*, its technical reviews of temperature and rainfall variability may be solid, but they neither contain any real opening for the governance challenges involved, nor illuminate what the real problems are that needs to be solved (assuming these are of a kind where governance, negotiations and human resources are needed). *Secondly*, even if it did identify the problems and their governance requirements correctly, MRC has no political clout or legitimacy to actually make the riparians act according to its findings. As above, and as the Xayaburi and the Don Sahong dams underline, there is a major possibility that the MRC will be sidelined when the discourse becomes politicised. Already, as we shall see below, the MRC and its CCAI are not in tune with, and seem to wield very little influence over, the national adaptation plans. The coming 'decentralisation' of the MRC mentioned above constitutes a development that should force the merger of regional and national agendas, although many expect that this also constitutes a de facto removal of a basin-governance perspective from the Mekong.

In the wider basin development strategies, GCC coincides (i.e. being discursively made to coincide) with political issues and their particular interests, in particular on how deeply regulated water allocation shall be and whether mainstream hydropower dams should be built or not (and how many, where, by whom, and for whom). As it seems, the conservationist and environmentalists argue that with GCC, uncertainties and risks have been added and therefore major interventions in the natural flow regime must be limited to a minimum, hence dam-building is to be avoided until knowledge is consolidated. However, dam-builders and modernists turn the argument on its head and argue that since there are new uncertainties regarding water flow, the only way to avoid future disasters is to regulate the flow through extensive dam-building (Interview, senior expert, Vientiane 2012). Only this way can wet season flow be reduced/controlled and dry season flow enhanced. The various actors' positions remain the same as before GCC, and it is at large exploited in order to support existing positions and interests.

The MRC, at bottom reflecting the aggregated interests of the riparians, takes a more conciliatory position in the infected hydropower debate:

> While climate change will affect the river's flow regime, the potential impacts of hydropower and irrigation developments also need to be considered in prediction of the future flow regimes . . . This creates complicated linkages between the hydraulic and ecological processes.
>
> (MRC 2009b: 2)

It does, however, which could be expected, lean towards the modernist argument on the benefits of regulation and interventions:

> The storage capacity of hydropower installations can therefore potentially reduce impacts of flooding on some areas. There is, however, a high degree of uncertainty and variation by location and between the different development and climate change scenarios.
>
> (MRC 2011: 4)

Rather than engaging politically through the MRC, the actions taken by the riparians are instead primarily visible, as we shall see, in the national programs. These are, as we also shall see below done prior to and irrespective of, the results of the MRC and the CCAI, based on domestic processes and rationalities. MRC and it member countries aimed for *adaptation* to the GCC, reaping the benefits from the international preparedness to fund projects. However, negligence and inability from NRC and riparians have rendered the CCAI impotent, making it appear more as a *subversion* of the real issues involved. The general problems that Dore & Lazarus identified above – the inability of MRC to position itself as a crucial actor in the basin governance matrix, and that the riparians act irrespective of MRC policies – are highly relevant here as well. Possibly it is even more acute in this field since GCC is both incremental and diffuse, making it easy to 'shrug off' for partners who do not agree with measures that need to be taken. Let us now scale down to national level and see what is occurring on that level, by the riparian actors.

Dealing with GCC at national level in the Mekong

Although the MRC estimates that there are approximately 300 projects being pursued throughout the basin in the name of adapting to GCC, there are also wide gaps in the national plethora of adaptations that need to be filled (MRC 2009: 4–5), and it appears clear that '. . . they are largely built on sectoral approaches and many of them suffer from the lack of coordination between different actors' (TKK & SEA-START RC 2009: 2). This is further emphasised by Kranz, Menniken, and Hinkel stating that national adaptation strategies have 'apparently been written in relative ignorance about each other's existence, displaying a lack of cross-referencing, transboundary analysis or planning, or the examination of potential synergies' (2010: 656). Having said that, on the project and policy level, all countries in the lower basin have been active, created national policies, and ratified the United Nations Framework Convention on Climate Change (UNFCCC) and the Kyoto Protocol (which Kranz et al. also is emphasising). Let us briefly review the content and institutionalisation of GCC-adaptation in each riparian country.

Vietnam has adopted its 'National Target Program to Respond to Climate Change' (2008) with the 'Ministry of Natural Resources and Environment' as the focal point. The program emphasises the risk for sea level rise, and the future inability to produce sufficient food for its citizens. Projections say that it could affect GDP with up to 10 per cent nationally by 2100, and up to 25 per cent in the Delta region. Especially, the low-lying Mekong delta is expected to be hard hit by a projected 10 metre sea level rise, from which massive relocation of people would follow. Consequently, it is also concerned by the possibility of lower flow in the Mekong (and Red River) mainstream.[10] In addition to the incremental reduction of farmland and available water resources, Vietnam is also concerned by the disaster aspects of GCC:

> Climate change is likely to increase the frequency, intensity, oscillation and extreme levels of dangerous weather phenomena such as storms, whirlwinds or temperature and rain induced natural disasters such as dry weather condition, floods, or droughts, damaging cold, salt intrusion, pests, which may all result in decreased livestock and crops productivity.
>
> (MoNRE 2008:12)

The National Target Programme (NTP) is to be implanted by a certain multi-ministerial organisation chaired by the Prime Minister, and with the Ministry of Natural Resources and environment (MoNRE) as the standing vice chair and lead ministry, with international experts in an advisory role (MoNRE 2008: 44).

Thailand's MoNRE emphasises that it is sensitive to climate change, especially through the enhanced probability of severe floodings,[11] prolonged droughts, and long-term reduced fresh water availability (MoNRE, Thai, n.d.). The Thai *Action Plan on National Climate Change as the Five Year Strategy on Climate Change 2008 to 2012* was the key document directing its efforts at preparation and adaptation. It stresses capacity building, understanding, and cooperation, or in the words of a national assessment study:

> The emphasis is on capacity-building, from research through public aware-ness and in the bureaucracy. Greater emphasis also needs to be placed on cooperation among government ministries and departments as there is a tendency to compartmentalize climate change analyses and reactions, whereas most research-based evidence underlines its' cross-cutting nature.
>
> (Lebel et al. 2010: 3)

Thailand may be the country in the basin which is the least oriented towards the Mekong basin water governance in its preparations for the GCC. It has however, signed and/or worked with all the key agreements/actors in the field (from Kyoto to UNFCCC; from Singapore Declaration on Climate Change to ASEAN's Climate Change Initiative) it has hosted a number of high-profiled climate meetings, and there is a wide variety of ministries involved in the capacity-building of the general GCC-preparations (Lebel et al. 2010: 4ff). It has a high ambition of mainstreaming the work of regular ministries, and may have reached a more sophisticated localised understanding (Chinvanno & Kerdsuk 2013), and/or has a lower dependence on external funders. It is, nevertheless, clear that it shares with all other countries in the lower basin the need of enhancing experts' as well as the public's knowledge on the issue.

In Cambodia and Lao PDR, the formalised responses are to be found in the National Adaptation Program of Action to Climate Change (NAPA). *Cambodia* is, like Vietnam, highly vulnerable to changes in the water regime. While Vietnam has to consider implications for the delta, Cambodia has the very special – and very productive for both agriculture and fisheries – water regime of Tonle Sap Lake to consider, which requires certain flows in the mainstream.[12] It is also aware of the stakes involved at the very top political level:

> Cambodia is highly vulnerable to climate change, the more so as it has low adaptive capacity to changing climate conditions. In recent years, we have witnessed more frequent and severe floods and droughts, which have resulted in a significant number of fatalities and considerable economic losses.
>
> (Stated by PM Hun Sen and quoted in RGC 2006: i)

The Cambodian NAPA is hosted by the Ministry of Environment and aims to: 1) understand the main characteristics of climate hazards in Cambodia; 2) understand coping mechanisms at the grassroots level; 3) understand existing programmes and institutional arrangements; and, 4) identify and prioritise adaptation activities (RGC 2006). While the Cambodian government has been working with UNFCCC, GEF, and UNDP on these issues, the government is also clear that there are severe barriers to implement this action plan, such as limited technical capacity with local and central government, inadequate integration of various programs, as well as limited general awareness of the climate change (RGC 2006). In addition, the institutional situation has proven to not be entirely clear when project financing started to pour in, and rivalry among stakeholders emerged (Interview, senior policymaker, Phnom Penh, 2012).

Lao PDR – as the other country in focus in this chapter – is sensitive to changes in the current climate regime. UNDP Resident Representative in Vientiane expressed it like this:

> With over 70% of Lao people depending on natural resources for their livelihoods . . . the people of Lao PDR are highly susceptible to the impacts of climate change. Predicted increases in the intensity and frequency of extreme climatic events will affect agricultural production, water supply, household food security, and the dynamics of water- and vector-borne diseases.
>
> (Rana, quoted in WREA 2009: 2)

It has, likewise, produced a NAPA in which it displays the awareness and recognises the need for adaptation and enhanced preparation. 'In this connection, the Government of the Lao PDR endeavours to find practical solutions to the challenges posed by climate change at a national level' (Lao PM Asang, quoted in WREA 2009).

Vulnerability is as high in Laos as in the other LMB countries. In the words of ADB:

> The Asian Development Bank (ADB) predicts that increasing temperatures and rainfall variability throughout Southeast Asia will cause a 2.5–10 per cent decrease in crop yields by 2020 and a 5–30 per cent decrease in crop yields by 2050.
>
> (ADB 2007, taken from WREA 2009: 8)

The key objective of the NAPA in Lao PDR is to deal with issues of urgency in relation to GCC. Key sectors identified for action are agriculture, forestry, water and water resources, and human health. The barriers to implementation of the NAPA are almost identical to those identified in the Cambodian case, namely insufficient base data, weak institutional capacity and low level of public awareness of the meaning and consequences of GCC (WREA 2009). In addition, in the Laos NAPA, objectives are wide and general, and there seems to be a vague assignment of responsibility for the implementation of the (WREA 2009).

Across the four riparians in the lower basin, the pattern is that nationally based measures are thinly integrated into in the wider national development strategies,

'In general, climate change issues are not integrated into the broader policy frameworks of national Governments or in specific sector and local government development plans' (MRC/ICEM 2010: 164), and even less so in the regional plans and policies. Moreover, the relation between GCC-adaptation at national and basin level are marginal. For instance, in the program developed in Vietnam and that of Thailand, 'MRC' is not mentioned at all, in the Cambodian NAPA only once, and in the Lao version only as reference for data. All documents are entirely without words such as 'politics' and 'governance', seemingly reflecting an unwillingness to enter into the political realm. They reflect very little 'basin-thinking' and even less any recognition of the need for mutual basin co-ordination (cf. Kranz et al. 2010). In contrast, the various documents tend to emphasise 'understanding', 'capacity building', 'projects', 'communication', and 'knowledge'. Hence, key ambitions typically stop at either localised projects or at 'soft' interventions. While there also are frequent mentioning of inter-ministerial cooperation and integration into national plans, few measures are actually presented to this cause.

China and Myanmar presents slightly different dynamics; the latter have only marginal relevance for the Mekong basin, and has virtually not started its work on preparation for GCC, being in a phase of intensive political change. *China* is huge and experiences climate change in all its forms, ranging from sea level rise to excessively melting glaciers, and from drought to floodings (Farber 2011). There is a growing awareness at central policy levels of the serious impact it may have: 'Climate change will also produce far-reaching impacts on society, economy and other fields, and cause huge losses to the national economy' (State Council of PRC, 2008: 10), and that it will constitute 'threats to the safety of life and property, and to the normal order and stability of social life' (10).

China launched its first national policy plan on adaptation already in 2007, named 'China National Climate Change Programme', which was followed up in 2008 and in 2009 with implementation plans and a Progress report, dubbing efforts at adaptation 'present and imminent' (Farber 2011: 370). These reports received general praise from outsiders. WRI assessed them in a major report in 2009 as:

> The plan stresses the need for adaptation of human and natural systems without hindering economic development. There is also a great focus on national level policy/ legislative approaches to enhance China's overall adaptive capacity. The adaptation strategies proposed tend to be large in scope and scale – the vision of the plan is impressive, but the document lacks specific targets and action-steps for realizing these goals.
>
> (WRI 2009)

Moreover, at the Chinese Party Plenum in 2010 (which is a politically authoritative event), the Communist Party emphasised its commitment to GCC-adaptation in a stronger way then before:

> As parts of efforts for climate adaptation, China will emphasize capacity building to address extreme weather events . . . Scientific research on climate change will be strengthened, and R&D on and application of low

carbon technologies will be expedited . . . China will continue to insist the principle of common by differentiated responsibility, and will aggressively participate in international cooperation on climate change issues.

(http://www.chinafaqs.org/blog-posts/chinas-party-plenum-recommends-climate-actions-12th-five-year-plan, accessed 14 October 2013)

While strong statements, they have not yet been followed up as anticipated. Overall, adaptation activities are seen as being in their 'infancy' while the 'groundwork is being laid' (Farber 2011: 369; 373).

As regards water, there are mainly engineering adaptation solutions foreseen, such as building of infrastructure to transport water and constructing more hydropower sites, whereas policy-wise, further coordinated and centralised planning is called for. At this stage, there is no separate plan being made (or at least made available) regarding the Yunnan province and the Lancang basin, although attention is being called to melting glaciers, which is highly relevant for this river. Other Chinese sources highlight that the climate adaptation activities are 'particularly [relevant for] the case of the Lancang river basin in China' (Wu et al. 2012: 427).

In the above, there is no indication that China has a dimension of transboundary concerns for the consequences of adaptation (or the lack thereof). To the contrary, interventions on the Lancang seem to be subject to entirely different rationales (cf. Magee 2006). If anything, the adaptation dynamics tend to rush hydropower construction and water deviation projects which historically has been unilateral Chinese ventures, in opposition to transboundary cooperation (Magee 2012).

Most of the basin countries recognise and act upon the looming GCC and the risks it entails. But they do so in the narrow frame of their own planning horizon, with scant, if any, reference to the regional, the common, good. Again, by default, at the end of the day, the reaction to GCC appears as *subversion* rather than *adaptation*.

Conclusions – to what extent is GCC adapted to, resisted, and/or subverted in GCC in the Mekong basin?

To what extent, and how, are the key actors and institutions responding to the emerging GCC? As in all the empirical chapters of this volume, we use the concepts of *adapt*, *resist*, and *subvert* to give some key characteristics of the process at hand.

Adapting

The region and its key actors have willingly accepted the projections of the IPCC and others, and are preparing themselves to take actions. Many minor projects and non-committing action plans are pushed forward, minor institutional adjustments are pursued, and funding is available (mostly technical ones for enhancing understanding). However, through weak citizen engagement, thin popular discourse, and its non-politicised nature, not much structural change is visible. Moreover, the basin dialogue on GCC is almost absent – except in very protected technical fora – subdued by animosity over other issues, preferences for avoiding controversial political processes in the MRC, as well as the traffic jam of major processes and projects already occupying the main stage of the Mekong governance.

Resisting

There are very few signs of explicitly resisting the idea of GCC in the Mekong basin. Quite to the contrary, as we have seen above, it has rather been embraced and become the source for institution-building and awareness-raising. The only resembling resistance that is detected is sluggishness and low prioritisation, resulting in a certain form of neglect and absence of preparedness to act decisively. Resistance occurs, however, indirectly through deliberately sluggish approaches and via the avoidance of addressing key issues – what Scott (1987) labelled 'everyday resistance' – but this possibly relates better to the next category.

Subversion

While adaptation is initiated and outright resistance is minor, there are neverthe-less impediments to implementation of preparations. In spite of half a decade (or more) of activities and projects, there are only minor political processes at work, and although there is major program initiated at basin level, it has not triggered any major deviations from business as usual. In fact, the key debates (as reviewed above) are unmoved and rather the GCC dynamics has been used as arguments in these debates (notably the dam-building controversy). Also at the national level, we see how support for GCC adaptation has been subverted into national rivalries and become a part of the 'regular' struggle for resources, influence, and status. At bottom, no major political GCC-adaptation advocate has so far risen in the region, and the issue has not taken place on the core political agenda yet. In this sense, *subversion* may be the most appropriate way of describing the process of basin adaptation to GCC.

From the above we can conclude that responses so far to GCC in the Mekong basin appear paradoxical on a number of accounts. *First*, the impact is predicted to be massive, with what appears to be huge stakes involved, including major political, social and economic consequences. Yet, neither the public nor top politicians are particularly engaged. This may be explained by the combina-tion of GCC appearing elusive to the regular citizen, and that all regimes in the basin are authoritarian or semi-authoritarian, impeding a popular discourse of a critical nature to take root. Following Scott's analysis of the Southeast Asian political culture (1976), the peasantry will be docile up to a certain point of dif-ficulties to sustain livelihoods, after which resistance will be considerable (and possibly violent). Most likely, given the depth of the projected GCC-impact in the basin, the process will be politicised with regional implications, but only in the medium- to long-term future; in the short term, GCC is second to more intense debates.

Second, the comparative advantage of the Mekong basin governance – hence, ability to adapt to GCC – compared to many other basins is that there is a long-established and institutionalised basin cooperation, including a secretariat with acknowledged technical skills and a sophisticated political dialogue mechanism. Still, the most tangible responses to GCC are to be found on the national level throughout the basin, but with little regional coordination. This may be a rather predictable result of the difficulties of acting from the basin level, and that project funding most easily rests with the national level ministry system, and their eagerness for managing projects.

Third, in spite of GCC's significance, responses seem to drown either in more contemporary debates on dam-building or water allocation, or be disregarded in the face of more-urgent poverty problems. The paradox here – which most would recognise – is that neglect to face GCC will enhance that original problem. While problematic, this resonates well with the higher share of private capital and short-term profits that increasingly are driving the political economy of the region in general, and the dynamics of interventions in the basin in particular. It also resonates with this particular time in the historical evolution of the basin, where the construction of the mainstream dams seems to be about to start and controversies are crowding out each other. There is also a poverty trap involved here, where the short-term imperative of reducing poverty may be consuming abilities to pursue sustainable development and thereby long-term poverty reduction.

Fourth, donor funds are forthcoming, projects are created, and institutional adaptation is pursued at national as well as basin level. In spite of this, basin-level interventions are limited and nobody seems to believe it will make an immediate difference. At national level, severe critique is delivered against the vagueness of projects and institutions, while at the same time rivalry among ministries on who is to 'own' the GCC at national level is reported (senior civil servant, Phnom Penh). This may, at bottom, reflect the universally shared ignorance (and therefore emerging frustration) on how to best adapt to a future climate change; if one does not know what is coming, it is hard to make interventions in order to prevent the unknown.

Elusiveness and incrementalism in combination with inept political climate and structural poverty issues seems to make GCC hard to address with proper governance tools. The inherited transboundary aspect in basin governance make matters even more complicated. So in spite of professional awareness, urgent needs, and available funding, responses on basin level are lagging.

Finally, the above may be exactly what can be expected. Just as the impact of GCC is incremental, the measures to adapt to it will be. Overall, also the most precise technical research from the natural sciences are elusive in degree, location, and time, and the most advanced social sciences are unclear on how to best prevent the harmful effect for basins, states, and communities. The best one can hope for at this stage is that there is a preparedness to deal with issues as they are identified. Basin governance has at least taken one first step in this direction.

Notes

1 This argument is equally relevant for the upper reaches of the river basin, in the Chinese parts of the basin named Lancang (Wu et al. 2012).
2 'The existing legal and, institutional framework in most Asian countries remains, inadequate to facilitate implementation of comprehensive and, integrated response to climate change in synergy with the pursuit, of sectoral development goals' (Cruz 2007: 493).
3 The river originates in Tibet, China and flows for a long stretch through China where it also falls rapidly. It touches upon Myanmar, before it enters Laos, constituting a border-river to Thailand for a while, returning into Laos and then entering Cambodia. In Cambodia, the rapid flow slows as it enters the plains before reaching the delta, southeast of Phnom Penh. The major share of the delta is in the southernmost part of Vietnam. The upper part of the river, situated in China, is called Lancang; the full name of the river basin is, then, the Lancang-Mekong river basin.
4 See Johnston et al. 2010 for a comprehensive summary of existing reports and their claims.
5 This brief review is based on a screening of the major English-speaking dailies/weeklies in the region since 2008, comprising *Bangkok Post*, *The Nation*, *Vientiane Post*, *Phnom Penh Post*, and *South China Morning Post*. Not all channels were fully accessible, nor all had archives covering the full time-period. I am grateful to Malin Hasselskog for assisting with this screening.
6 In the most recent USAID-sponsored report (2013), eight 'hot-spots' were mentioned: Chiang Rai, Sakon Nakhon in Thailand; Kahammouan and Champasak in Laos; Stung Treng, Mondulkiri and Kampong Thom in Cambodia; and Kien Giang and Gia Lai in Southern Vietnam.
7 It should also be mentioned that Myanmar is a riparian to the Mekong basin. It plays, however, no significant role for the moment and it has no major plans for withdrawing any major amounts of water from the basin.
8 The MRC sees itself to be 'well placed to assist member states' on: developing policy frameworks; making integrated assessments of impacts and vulnerability; reviewing adaptation options; making pilot project for demonstration effects; engaging stakeholders; monitoring and reporting on progress; and building capacity and awareness (MRC, CCAI pamphlet, n.d.).
9 'The project focuses on identifying the environmental, economic and social effects of climate change in the Lower Mekong Basin (LMB), and on assisting highly exposed and vulnerable rural populations in ecologically sensitive areas increase their ability to adapt to climate change impacts on water resources, agricultural and aquatic systems, livestock, ecosystems, and livelihood options' (USAID 2013: 4).
10 Below a certain dry season flow, the fresh water is unable to hold back the pressure from sea-water intrusion, which already now can be measured all the way to Phnom Penh. With a rising sea level, and a lowered dry season flow, the effect would be critical.
11 The massive rainfall in the summer/autumn of 2011 in Thailand and the ensuing 'slow flooding' eventually engulfing large parts of Bangkok Metropolitan area, and submerging a vast number of industrial estates, having implications on global scale, served to convince Thai authorities on its vulnerability for water-related extreme events (Thanasupsin 2012).
12 The MRC-agreement is a framework which does not state exact figures on water flow and allocations. It does, nevertheless, recognise, as the only particular site, the necessity of preserving the current regime around Tonle Sap, its importance for Cambodia, and for the entire basin (MRC 1995).

References

Apichai Sunchindah and Petra Pailin Mueller 8 October 2009. ASEAN is at the forefront of the Climate Change Challenge. The Nation. Retrieved 1 March 2015 www.nationmultimedia. com/home/ASEAN-is-at-the-forefront-of-the-Climate-Change-Ch-30114002.html

Bangkok Post 9 April 2013. Mekong study predicts crop shifts. Bangkok Post. Retrieved 1 March 2015 m.bangkokpost.com/news/344619

CCAI, 2010, MRC, Vientiane. http://www.mrcmekong.org/about-the-mrc/programmes/climate-change-and-adaptation-initiative/

Chinvanno, S., and Kerdsuk, V., 2013, *Mainstreaming Climate Change into Community Development Strategies and Plans: A Case Study in Thailand, Adaptation Knowledge Platform*, Partner Report Series No. 5. Stockholm Environment Institute, Bangkok, accessed 16 February 2015 at http://www.climateadapt.asia/upload/publications/files/5121ba0bdf95eThailand_Web.pdf

Cruz, R. V. et al., 2007, 'Asia: climate change 2007: impacts, adaptation and vulnerability. Contribution of working group II to the fourth assessment report of the intergovernmental panel on climate change', in Parry, M. L. et al. (eds.), *Summary for Policymakers, in Climate Change 2007: Impacts, Adaptation and Vulnerability. Contribution of Working Group II to the Fourth Assessment Report of the Intergovernmental Panel on Climate Change*, 469–506. Cambridge: Cambridge University Press: 469–506.

Danny Marks. 30 August 2010. Thailand must take climate change seriously. The Nation. Retrieved 1 March 2015 www.nationmultimedia.com/others/Thailand-must-take-climate-change-seriously-30136846.html

David Montero 6 June 2008 In Cambodia, a case for localizing climate-change research. The Christian Science Monitor. Retrieved 1 March 2015 www.csmonitor.com/Environment/Global-Warming/2008/0606/in-cambodia-a-case-for-localizing-climate-change-research

De Stefano, L., Duncan, J., Dinar, S., Stahl, K., Strzepek, K., and Wolf, A., 2012, 'Climate change and the institutional resilience of international river basins', in *Journal of Peace Research*, 49(1): 193–209. doi: 10.1177/0022343311427416.

Dore J., and Lazarus, K., 2009, 'De-marginalizing the Mekong River commission', in Molle, F., Foran, T., and Käkönen, M. (eds.), *Contested Waterscapes in the Mekong Region: Hydropower, Livelihoods and Governance*. London: Earthscan: 357–382.

Drieschova, Alena, Giordanoa, Mark, and Fischhendler, Itay, 2008, 'Governance mechanisms to address flow variability in water treaties', in *Global Environmental Change*, 18: 285–295.

Eastham J., Mpelasoka, F., Mainuddin, M., Ticehurst, C. C., Dyce, P., Hodgson, G., Ali, R., and Kirby, M., 2008, *Mekong River Basin Water Resources Assessment: Impacts of Climate Change*. CSIRO: Water for a Healthy Country National Research Flagship, accessed 16 February at http://wacc.edu.vn/vi/wp-content/uploads/2013/06/wfhc-MekongWaterResources-Assessment.pdf

Farber, Daniel A., 2011, 'The challenge of climate change adaptation: learning from national planning efforts in Britain, China, and the United States', Research Roundtable on Climate Change, Adaptation, and Environmental Law, 7–18 April 2011, University of California, Berkeley.

Fox, Colleen, and Sneddon, Chris, 2005, 'Flood pulses, international watercourse law, and common pool resources: a case study of the Mekong lowlands', WIDER Research Paper 2005/22, Helsinki.

Hang, Pham Thanh, and Lennaerts, Ton, 2008, 'Joint basin development planning to foster IWRM in the Lower Mekong basin', paper presented to the Symposium, *Mekong Management at a Watershed – IWRM in the Global Crisis*. Gothenburg University, Sweden.

Gerlak, A. K., and Schmeier, S., 2014, 'Climate change and transboundary waters: a study of discourse in the Mekong River commission', *The Journal of Environment & Development*, 23(3), 358–386.

Hirsch P., and Mœrck-Jensen, K., 2006, *National Interests and Transboundary Water Governance in the Mekong*. AMRC: University of Sydney.

Hirsch, P., 2010, 'The changing political dynamics of dam building on the Mekong', in *Water Alternatives*, 3(2): 312–323.

Houghton, J. T., et al. (eds.), 2001, *Climate Change 2001: The Scientific Basis. Contribution of Working Group 1 to the Third Assessment Report of the Intergovernmental Panel on Climate Change*. Cambridge and New York: Cambridge University Press.

Jacobs, J. W., 1995, 'Mekong committee – history and lessons for river basin development', *Geographic Journal*, 161(2):135–148.

Johnston, Robyn, Lacombe, Guillaume, Hoanh, Chu Thai, Noble, Andrew, Pavelic, Paul, Smakhtin, Vladimir, . . . Sze, Choo Poh, 2010, *Climate Change, Water and Agriculture in the Greater Mekong Subregion*, IWMI Research Report No 136, IWMI, Colombo.

Keskinen, Markko, et al., 2008, 'Water resources development and impact assessment in the Mekong basin: which way to go?' *Ambio* 37(3):193–198.

Kranz, Nicole, Menniken, Timo, and Hinkel, Jochen, 2010, 'Climate change adaptation strategies in the Mekong and Orange-Senqu basins: What determines the state-of-play?' *Environmental Science and Policy*, 13: 648–659.

Kuroda, Haruhiko, and Stern, Nicholas, 5 May 2009, 'Why global warming could make or break South-East Asia', *The Guardian*, London, accessed 16 February 2015 at http://www. adb.org/news/op-ed/why-global-warming-could-make-or-break-south-east-asia

Lacombe. G., Smakhtin, V., and Hoanh, C. T., 2013, 'Wetting tendencies in the Central Mekong basin consistent with climate change-induced atmosphere disturbances already observed in East Asia', *Theoretical and Applied Climatology*, 111: 251–263.

Lebel, Louis, 2010, *Scoping Assessment for National Implementation, in Thailand, Summary*, AIT UNEP RRC, AP 2010, Bangkok.

Magee, Darrin, 2006, 'Powershed politics: Yunnan hydropower under great western development', *The China Quarterly*, 185, 23–41.

Magee, Darrin, 2012, 'The dragon upstream: China's role in Lancang-Mekong development', in Öjendal, Joakim, Hansson, Stina, and Hellberg, Sofie (eds.), 2012, *Politics and Development in a Transboundary Watershed – The Case of the Lower Mekong Region*. London: Springer: 171–193.

MRC, 1995, *Agreement on the Cooperation for the Sustainable Development of the Mekong River Basin*. Chiang Rai, Thailand: MRC.

MRC, 2009a, *Adaptation to Climate Change in the Countries of the Lower Mekong Basin: Regional Synthesis Report*, MRC Technical Paper No 24. Vientiane: MRC.

MRC, 2009b, *Adaptation to Climate Change in the Countries of the Lower Mekong Basin*, MRC Management Information Booklet Series No.1. Vientiane: MRC.

MRC, 2010a, *Impacts of Climate Change and Development on Mekong Flow Regimes – First Assessment 2009*, MRC Technical Paper No 29. Vientiane: MRC.

MRC, 2010b, *Review of Climate Change Adaptation and Tools*, MRC Technical Paper No 34. Vientiane: MRC.

MRC, 2010c, *State of the Basin Report 2010*, Mekong River Commission. Vientiane: MRC.

MRC, 2011, *Impacts of Climate Change and Development on Mekong Flow Regimes: First Assessment – 2009*, MRC Management Information Booklet Series No. 4. Vientiane: MRC.

MRC/ICEM, 2010, *MRC SEA for Hydropower on the Mekong Mainstream – Summary Baseline Assessment Report*. Vientiane: MRC, accessed 16 February 2015 http://www.icem.com.au/ documents/envassessment/mrc_sea_hp/2.%20baseline/reports/pdf/SEA_BAR_vol1.pdf

MS (Mekong Secretariat), 1970, *Indicative Plan for Hydropower in the Mekong River Basin*. Bangkok: Mekong Secretariat.

Öjendal, Joakim, 2000, *Sharing the good – modes of managing water resources in the lower Mekong River basin*, PhD dissertation, Department of Peace and Development Studies, Göteborg University.

Öjendal, Joakim, Hansson, Stina, and Hellberg, Sofie (eds.), 2012, *Politics and Development in a Transboundary Watershed – The Case of the Lower Mekong Region.* London: Springer.

Öjendal, Joakim, and Mørck Jensen, Kurt, 2012, 'Politics and development of the Mekong River basin – transboundary dilemmas and participatory ambitions', in Öjendal, Joakim, Hansson, Stina, and Hellberg, Sofie (eds.), *Politics and Development in a Transboundary Watershed – The Case of the Lower Mekong Region.* London: Springer.

Peter N. Spotts. 15 November 2007. How to fight a rising sea. The Christian Sceince Monitor. Retrieved 1 March 2015. www.csmonitor.com/2007/1115/p13s02-wogi.html

Pongphon Sarnsamak 2 April 2010. Mekong power plan will affect millions of lives: activists. The Nation. Retrieved 1 March 2015. www.nationmultimedia.com/home/2010/04/02/national/Mekong-power-plan-will-affect-millions-of-lives-ac-30126155.html

RGC, 2006, *National Adaptation Programme of Action to Climate Change (NAPA),* Ministry of Environment, Phnom Penh.

Scott, J. C. (1976). The moral economy of the peasant: Subsistence and rebellion in Southeast Asia. Yale UP (New Haven).

Scott, J. C. (1987). Weapons of the weak: Everyday forms of peasant resistance. Yale University Press.

Shackley, S., and Wynne, B., 1996, 'Representing uncertainty in global climate change science and policy: boundary-ordering devices and authority', *Science, Technology & Human Values,* 21(3), 275–302.

Shailendra Yashwant 23 January 2009 Developed countries have a climate-protection obligation. The Nation. Retrieved 1 March 2015 www.nationmultimedia.com/2009/01/23/opinion/opinion_30093941.php

State Council of PRC, 2008, 'China's policies and actions for addressing climate change', Information Office of the State Council of the People's Republic of China, Beijing.

Thanasupsin, S. P., 2012, 'Climate change impacts on water resources: key challenges to Thailand CC adaptation', accessed 16 February 2015 at http://www.rid.go.th/thaicid/_5_article/7symposium/7th-13.pdf

Thapan, Arjun, 2009, 'A 2020 vision for the Mekong's sub region', *The Nation,* 16 June 2009, Bangkok.

TKK & SEA-START RC, 2009, *Water and Climate Change in the Lower Mekong Basin: Diagnosis and Recommendations for Adaptation.* Helsinki and Bangkok: Water and Development Research Group & SEA START RC.

USAID, 2013, *Mekong Adaptation and Resilience to Climate Change.* Bangkok: DAI.

WREA, 2009, *National Adaptation Programme of Action to Climate Change.* Lao People's Democratic Republic, Vientiane.

WRI, 2009, 'National climate change strategies: comparative analysis of developing countries', accessed 15 September 2013 at http://pdf.wri.org/working_papers/developing_country_actions_table.pdf

Wu, Feifei, Wang, Xuan, Cai, Yanpeng, Yang, Zhifeng, and Li, Chunhui, 2012, 'Spatiotemporal analysis of temperature-variation patterns under climate change in the upper reach of Mekong River basin', *Science of the Total Environment* 427–428: 208–218.

WWF, 2009, *The Greater Mekong and Climate Change: Biodiversity, Ecosystem Services and Development at Risk.* Bangkok, Thailand: WWF.

Zelda Soriano 1 May 2013. Climate change demands that Asean's future is cooperative. The Nation. Retrieved 1 March 2015 www.nationmultimedia.com/opinion/Climate-change-demands-that-Aseans-future-is-coope-30205127.html

Zhu, Lijiang, 2011, 'Chinese practice in public international law: 2010', *Chinese Journal of International Law,* 10(2): 427–468, doi: 10.1093/chinesejil/jmr013.

5   The Niger River basin

Introduction

The Niger River is the second largest river in Africa by discharge volume (5,600M³/s at Onitsha; 1955–1991), and the third longest river in Africa (Grijsen et al. 2013). The basin involves nine countries,[1] all of which can be found at the lower end of the Human Development Index. Only Nigeria and Ivory Coast are not classified as least-developed countries (LDCs). Water, food, and energy are in short supply and important measures are needed in order to lift the region and its population out of an exceptionally precarious and vulnerable situation. In order to do that, most of the basin riparian states have identified agriculture as the engine of development, a sector that largely depends on water resources management. The predictions that the region will be hard hit by climate change, both in terms of temperatures and changing rain patterns, with severe effects on livelihood possibilities, are particularly threatening considering the very limited resources to deal with such effects.

The Niger is 4,200 km from its source in the Fouta Djallon Mountains in the south of Guinea[2] where it is joined by several tributaries, to its outflow in the Gulf of Guinée in Nigeria. On its way eastward, it covers a surface of 2.117.700 km² in nine countries. The mean annual flow varies significantly along the flow of the river. The annual mean flow into the inner delta from Guinea and Mali is 46km³, while it has decreased to only 33km³ at Taoussa in Mali in the upper bend of the river (Ogilvie et al. 2010: 599–600).

From Guinea, the Niger enters Mali to the north, where it is joined by an important tributary, the Bani River.[3] Between Segou and Timbuktu, the river opens up into the biologically important inner delta, also known as Macina.[4] The delta is important for cultivation as well as fisheries, and it provides the most important habitat for migrating birds in the region. In the inner delta, the river loses much of its flow to seepage and evaporation as it diverts into a complex network of channels, lakes, tributaries, and swamps. Two-thirds of the river's potential flow is reduced in the Macina. After the delta, the river forms the Niger loop as it touches the Sahara desert and takes a southward turn through the arid areas of Mali and Niger, where it further loses some of its flow. Despite the important inflow from tributaries upstream, the flow of the river is less when it enters Niger than at its source in Guinea. The river then constitutes the border between Benin and Niger before it enters Nigeria. In Nigeria, it is joined by several tributaries, the most important of which is the Benue River.[5] As the Niger River enters the Gulf of Guinea and the Atlantic Ocean it again opens up, this time as the oil-rich maritime Niger delta.

The Niger supports a population of around 125 million people (of which 70 per cent live in Nigeria), with a population growth of 2.63 on average/year.[6] The basin countries' populations to a large extent concentrate along the river and the population density in the basin is four to five times higher than the national averages (Ogilvie et al. 2010). A majority of the population (64 per cent) live in rural areas and are mainly relying on rainfed agriculture for livelihood. As much as 78 per cent of total agricultural production is subsistence agriculture. It is characterised by low input, short growing seasons, dry spells, and excessive water withdrawal. It consists of nomadic pastoralism in the north, and irrigated and fertilized agriculture further south. Fishing is practiced throughout. Poverty is prevalent in the region and much of the population suffer from 'extreme, chronic poverty and remains vulnerable to droughts and malnutrition' (Ogilvie et al. 2010: 594).

It is clear that people under these conditions are particularly vulnerable to climate change, but also that there is a desire to engage in large-scale poverty-reducing measures – measures that may stand in opposition to more long-term sustainability. Access to electricity is low in all countries but Nigeria (yet even in Nigeria demand exceeds supply (Shackley & Wynne 1996)), which motivates an extension of dams to produce 'clean' energy. Moreover, there is important irrigation potential, and need in a region that is heavily affected by irregular rainfall. West African rivers are fairly undeveloped in terms of major infrastructure as compared to other major water courses in Africa. More than 90 per cent of the flow pass unregulated (Barbier et al. 2009). At the same time, the river is threatened by soil erosion, deforestation, over exploitation and pollution, domestic, as well as industrial, agricultural, and mining. Effects that are being related to global climate change, such as rising temperatures, desertification, and more severe droughts as well as concentrated rains that cause floods, increase the concerns about the vulnerability of the populations in the basin area. Improving the lives of the population in the river basin through economic development while at the same time increasing adaptability and resilience to climate change thus poses an important challenge for transboundary water management (TWM) institutions.

However, cooperation in the basin is made difficult by recurring political instability. West Africa has been characterised as 'one of the world's most sensitive and volatile regions' (Mazzitelli 2007, in Obi 2008: 184), or the 'coup d'état belt' of Africa (Olukoshi 2001: 4). Since independence, several countries have experienced a number of coup d'états, military rule, and civil war. During the last decade, all riparian states except Burkina Faso and Benin have experienced violent conflict. The Guinea government was ousted in a coup in 2008, with the resulting exclusion of the country from ECOWAS and AU in 2009. Since then, elections have been held, but political violence has resurfaced. Ivory Coast saw a brutal civil war in the early 2000, and Niger experienced military coups with ensuing elections in 1999 and 2010, as well as a second Tuareg rebellion in between. Nigeria is facing violent opposition both from different groups in the Niger delta, and by the Islamist group Boko Haram in the north. In 2012, violent conflict erupted in relatively stable Mali, when a Tuareg rebellion in combination with Islamist takeover of regions in the north was followed by a military coup. Although stability has since returned in Mali and elections have been held, the situation in the region continues to be volatile.

The high prevalence of poverty in the region has affected international relations. The major priority of foreign policy of many riparian states has been to seek cooperation outside the African continent rather than with their neighbours, closely linking foreign policy and cooperation policy (Mamadou 2001). Particularly, the francophone countries in the basin (all but Nigeria) have strongly focused on their relations with France. The divide between the former French colonies and the former British colonies has been considered an important obstacle to regional cooperation and integration (Rönnbäck 2008). Nigeria, as the hegemon in the region, has been an important promoter of regional integration across the language divide through the ECOWAS, partly with the purpose of limiting the influence of France in the subregion (Obi 2008). More importantly though, integration is motivated by an increasing awareness of 'the evolving linkages between national, subregional and transregional threats' (Obi 2008: 184)

Instability in the region has caused important refugee flows, an increase in transborder crimes and movements of small arms, as well as cooperation between extremist groups established in the region, for example, make the recent Mali conflict an urgent concern for Nigeria and its fight against local extremist groups. These spillover effects have increased awareness of interdependence in the sub region and have resulted in far reaching common security initiatives (Obi 2008). Despite such initiatives, the actual potential for regional cooperation and the role of the security mechanisms of ECOWAS have been questioned (Obi 2008). It has been argued that necessary political transformation and social justice stand in opposition to the interests of West African elites and their neo-patrimonial rule, which is based on personalised structures of authority. Despite institutionalised norms of democratic governance, it has been argued that the status quo is perpetuated by political leaders who sanction each other's efforts to maintain their grip on power (Taylor and Williams 2008: 146). A political culture of personalised rule and foreign relations based on individual connections also plays an important role in sub-regional cooperation in general (Obi 2008). As such, regional initiatives are highly dependent on the commitment of elite figures who manage to draw support from peers in the sub-region. Since the new millennium, TWM in the Niger River basin has benefited from such commitment in combination with the imperative to secure donor support through initiatives of regional cooperation and adoption of the climate change discourse. However, the way in which the riparians manage to cooperate over the management of the river, and the way in which the climate change discourse is played out in practice will be shaped by harsh competition over attention and resources in a region characterised by poverty and conflict.

Climate variability and climate change projections in the Niger basin

The climate in the Niger basin is characterized by a rainy season in the summer and a dry season in the winter/spring, caused by the seasonal movement of two air masses separated by the inter-tropical convergence zone (ITCZ). The climate ranges from humid equatorial at the coast to semi-arid and arid in the northern Sahelian countries. This means that annual rainfall varies from 1,500 mm/year in the tropical areas to 250 mm on the desert fringe, where temperatures range from 10–50 degrees. Due to the shape of the river and its northward-southward bend as it touches the Sahara desert in Mali, it passes through each climatic zone twice, which affects its hydrological cycle (Abrate et al. 2013).

The region is characterized by important climatic variations over time. Sediment studies have shown that the region has seen greater dry periods in earlier centuries (Barbier et al. 2009: 32), and that huge variability in climate is the norm (Goulden & Few 2011: 16). Monitoring of Niger River flow goes back to 1907 when the first station was installed at Koulikoro in Mali (Andersen et al. 2005). Measurements show that the climate in the region saw an important break in 1970, separating a humid period before, and a more recent dry period. In the period 1968–1997, there was a 29–49 per cent decrease in rainfall, compared to the 1931–1960 period, with variations in different areas (IPCC 2001, in Kandji et al. 2006: 9). During the same period, there was a delayed onset of the rain period and a decreasing trend in the duration of the wet season in some areas. The region was hit by severe droughts in 1968, 1973–1974, 1983–1985, and in 2005, and drier years occur regularly. In 1985, a zero-flow condition of the Niger River was observed in Niamey. There has also been a 200 km southward shift in isohyets, and the desert is advancing at a rate of 700 metres per year on average (Okpara et al. 2013). Earlier explanations for the decrease in rainfall have focused on 'over-cultivation', 'overgrazing', and excessive exploitation of wood fuel, but have more recently been complemented by research suggesting a strong link between sea surface temperature anomalies and rainfall in the Sahel region (Giannini et al. 2003; Olsson et al. 2005). According to the IPCC, it is difficult to determine whether the drying was a manifestation of long-term climate change or a protracted cycle.

The decrease in rainfall ended in 1988 with a return to 'normal' rainfall. During the last 20 years, the Sahel region has witnessed an increase in severe floods: in 1995, 1998–1999, 2002–2003, 2006–2008, 2010, and 2012 in some parts of the basin. They have been associated with factors such as unusual heating in the tropical Atlantic Ocean and La Niña in the tropical Pacific Ocean (Paeth et al. 2009). At the same time, the region has seen a consistent trend of increasing vegetation greenness that has been called the greening of Sahel, and a rising water level in some aquifers. The increasing rainfall does not seem to fully explain the changing vegetation pattern, and the water level rise, it seems. Other explanatory factors might be changed land management (grazing, cropping, manuring), new agricultural policies, and demographic trends (Olsson et al. 2005: 561).

Africa in general and the semi-arid and arid regions of West Africa in particular, are expected to be hit hard by climate change, not least when it comes to water stress, according to the IPCC. However, just as in the case of the Nile (see this volume), there is no agreement on whether West Africa will become drier or wetter in the near future.[7] Half of the models used by IPCC predict increased rainfall and the other half predict a decrease. Therefore the IPCC tool does not project changes in river discharge. What there seems to be agreement on is that temperatures will rise and variability will increase and extreme events exacerbate (Grijsen et al. 2013: 48; Okpara et al. 2013: 29). Rainfall patterns are expected to become more erratic in terms of quantity, timing, and geographic scope (RI 2013). Hence, any long-term predictions are very difficult to make. However, irrespective of the direction changes take, projections show important threat to food production in the long term (Kandji et al. 2006: 11).

Climate change is not necessarily seen as the greatest threat to the environment and people's livelihoods. Instead, a number of studies indicate that changes in human activity, such as population growth patterns, settlement, and irrigation, will have impacts that equal the predicted effects of climate change (Mahé & Paturel 2009; Kerres 2010).

In a context of important climatic variability, the population has for a long time had to adjust to uncertainties, and make their productive activities compatible with the environment, finding complementary sources of livelihood. The severe droughts of the past several decades have further pushed changes in agricultural practices (Mounir et al. 2011). Adaptive mechanisms developed within traditional agricultural systems are, for example, sowing in multiple fields, kilometres apart, as a risk-spreading strategy. Other examples are planting pits, installing micro-dams and catchments, and constructing soil ridges and stone lines to conserve soil and moisture (Greaf & Haigis 2001; Goulden & Few 2011: 22). Traditional mechanisms are also combined with modern techniques, for example through the use of mobile phones for diffusion of predictions that make assessment of environmental conditions in distant fields possible. However, for the poorest groups in society, the increasing pace with which extreme events occur makes it more difficult to rebuild their adaptive capacities in 'normal' years.

Increased agricultural output and energy production are prioritised areas in poverty reduction strategies throughout the region. For that purpose, exploitation of the river provides a potential for economic development in the whole region, particularly when it comes to irrigation and energy production, both of which are severely lacking. The irrigation potential for the whole basin is estimated at 2,816,510 ha, of which only 924,610 ha are already under irrigation. It is estimated that hydropower could generate 30,000 gigawatt hours per year, but only 6,000 have been developed so far (Grijsen et al. 2013). It is reasonable to assume that perceptions of climate change effects will contribute to shape investment plans and the possibility of producing sustainable effects on people's livelihoods.

GCC Discourse in the region

> Africa will be the continent hardest hit by climate change because it faces more severe climatic effects than other regions, its economies rely on climate-dependent sectors such as agriculture, and its capacities to cope and adapt are generally limited.
>
> (UNDP 2011)

As we saw above, it is difficult to establish a direction of climate change, as well as to what extent certain tangible effects such as desertification or changing rainfall patterns are a consequence of local activities or global climate change. Here the focus is on how climate change is discursively played out in the basin, with particular focus on the media and development assistance. As we saw, the overarching perception is that West Africa will be severely hit, and that climate change will add to an already harsh and vulnerable situation.

A scanning of major newspapers in the region shows that they connect a broad range of problems to climate change effects. The Nigerian daily newspaper *Vanguard* illustrates the certainty expressed in media as they state that climate change 'is now glaring for all to see' (Tope 2013). Climate change is portrayed as irreversible, and adaptations emerge as the way to address it. The media attention paid to climate change in the different riparian countries is quite similar. Only Nigeria differs somewhat as climate change is to a larger extent debated in opinion pieces, for example emphasising the global responsibility and blaming globalisation as in the *Daily Trust* in October 2013, where it is stated that 'environmental challenges such as climate change, cross-boundary water, air-pollution, and over fishing of the oceans are also linked with globalization. The truth is globalization is destroying everything' (Enahoro 2013).

Apart from that, a number of themes seem to be organising the media discourse on climate change. First and foremost, climate change is discussed in relation to extreme events such as droughts and floods. The population is suffering and climate change is portrayed as aggravating already serious and well-known threats and challenges in the region. Most particularly, food security and local economies and livelihoods are threatened.

Second, a number of links are made between climate change and other challenges, such as health, migration patterns, and conflict (specifically between herders and farmers, but also to the growth of terrorism such as Boko Haram, the Niger delta conflict, and the Mali conflict). The river Niger does not play a prominent role in those accounts. However, in case of flooding there is a tendency to blame climate change rather than, for example, siltation. Moreover, there are instances, for example in Nigeria, where floods have been made issues of national politics as the federal government has been accused of passivity in face of flood warnings, and provision of assistance (Maduforo et al. 2013). National institutions, in their turn, lament the difficulty of making people living close to the river react to early warnings (Elhadj 2012; Maduforo et al. 2013).

Third, science and the use of local knowledge emerge as solutions, and great faith is placed in them. It concerns, for example, the dissemination of new agricultural techniques and better adapted seeds, as well as conflict reducing measures such as grazing paths and conflict resolution mechanisms. Particular emphasis is often placed on the importance of local knowledge and tradition.

Fourth, there is much focus on initiatives and projects that address climate change adaptation. Workshops, sensitisation, capacity building, and information campaigns are noticed. External actors, INGOs, and donors often emerge as the ones initiating such action, but local farmers and NGOs also come through as active. Even politicians in the region emerge as committed and well informed. When Hama Amadou, president of the national assembly in Niger, travelled the country after the most recent flood in 2013, he came across as concerned and stated that although poverty of the country prevents the state from taking effective measures, the state would not abandon the population (Souleymane 2013). In Malian press, the state is shown to take action as it takes measures to provoke rain, and in both Mali and Niger the presidents are appreciated for committing themselves to making climate change a priority.

When regional cooperation is emphasised, it primarily concerns the Sahel region, rather than the river basin, and it takes the shape of activities performed by established regional initiatives, such as the Comité Permanent Inter-Etats de Lutte contre la Sécheresse au Sahel (CILSS), its research branch, Application en Agrométéorologie et Hydrologie Opérationelle (AGRHYMET), and donor initiatives that cover several countries. The Niger Basin Authority is largely absent from the media, except when its relaunch in the early 2000 was received with optimism, and the statement that good governance will save the Niger River (RFI 2004).

While actors in the riparian countries will be discussed and more thoroughly analysed in the latter part of this chapter, there are other actors whose activities in the region are crucial and merit some attention, namely actors who perform development assistance in different forms.

Major institutions such as UNDP and the World Bank that play an important part in poverty reduction in the respective countries emphasise the importance of climate change and climate change is also stated as the potentially greatest obstacle to the achievement of the Millennium Development Goals (MDGs) (Kandji et al. 2006). They, therefore, link their poverty reduction efforts to climate change adaptation, for example through the National Adaptation Plans, NAPAs. Considering the important role played by UNDP and the World Bank in the respective countries, their focus on climate change evidently focus resources and initiatives in that direction. However, reports commissioned by major actors tend to rely on scientific studies and are careful to point out that 'it is next to impossible to ascertain to what extent the greenhouse effect has had an impact on the decrease of rainfall and river flow from the early 1970s to the early 1990s, nor to what extent it has caused its rebound since then', as a report commissioned by the KFW states (Kandji et al. 2006; Kerres 2010;

Goulden & Few 2011). The 'normal' climate variability seems to be serious and unpredictable enough not to have to rely on climate change as a measure of urgency. When climate change is evoked, it is on top of 'normal' chocks, and is seen as aggravating an already precarious situation. The need for infrastructure to effectively mitigate the seasonal and annual variability of the Niger River is pointed out. However, it is also emphasised that improved operation and maintenance of the existing ones will ensure greater efficiency in the use of existing assets and ensure their long-term sustainability (World Bank 2007). As a result of the focus on variability and the uncertain direction of climate change effects, recommendations often focus on flexibility and avoiding investments that cause lock-in effects, and reduce capacity to adapt to either lower or higher water availability. No-regret measures, i.e. 'measures which are justifiable even in the complete absence of climate change and variability', are recommended (Kerres 2010: 23). Other issues that are more predictable, such as population growth in the region are taking precedence over climate change-induced risk. The impact of population increase is considered comparable to possible impact of climate change (Mahé & Paturel 2009; Kerres 2010).

While the river Niger is largely absent from the media discourse on climate change, it is very present in the development discourse and is an object of many initiatives and investments – especially when it comes to the relaunch of the Niger Basin Authority. In the following, we will look more closely at the history and development of TWM and the NBA, before we analyse current strategies of the riparian countries concerning climate change and transboundary cooperation in the shape of NBA, and other institutions.

Institutional history/development

Relations between the countries along the stretch of the Niger River basin have been characterised more by cooperation than by conflict, but with limited resources. During colonialism, in the early 1950s, an organisation to manage the water resources of the Niger River was created in Bamako. La Mission d'Etude et d'Aménagement du Niger (MEAN) had the upper part of the river, from Guinea to the Niger/Nigeria border (also the border between French and British rule) as its jurisdiction. The goal was to elaborate a master plan for the management of Niger, focused on the inner delta in Mali. However, with independence in 1960, the organisation was discontinued.

In 1963, the riparian countries instead adopted 'L'Acte de Niamey' concerning navigation and economic cooperation, recognising the international character of the river and the necessity to create joint regulation (NBA History). Soon after, in November 1964, the riparian states created the Niger River Commission (NRC), 'to encourage, promote and coordinate studies and programmes in relation to work to make use of the basin resources'. The commission is not considered to have succeeded in achieving much, and after 17 years it was replaced by another institution, the Niger Basin Authority, NBA, with a more encompassing task. The World Bank's review of the activities of the NRC states that '[u]nfortunately, the action plan was stillborn mainly because the enthusiasm of the donors in the mid-1970s built it up to a size that was quite beyond the counterpart implementation capacity of the then NRC or indeed of the member states themselves' (Rangeley et al. 1994).

In November 1980, the Niger River Basin Authority (NBA) was created. The nine member states include Burkina Faso, Benin, Cameroon, Ivory Coast, Guinea, Mali, Niger, Nigeria, and Chad.[8] The organisation faced an institutional crisis already in 1984, due to lack of financial support and lack of response from donors, as well as political turmoil in the member countries. The crisis lasted for 10 years and caused great damage to the development of the organisation. To face the crisis, the fifth meeting of heads of states in 1987 decided to concentrate the targets of the organisation and create an executive secretariat as well as focal points in the member countries. These reforms where consistent with the recommendations of a review made by UNDP and FAO, including a proposed five-year plan and a formula for funding (Rangeley 1994).

The UNDP, the World Bank, ADB, and several bilateral donors (including France, Germany, Canada, the United States, and the Netherlands) have supported the Niger Basin Authority; however, with limited result. Limitations have been presenting themselves in terms of insufficient budgetary support from the member countries, constraints to the planning process, the difficulty in implementing regional projects in a multinational political setting, poorly defined (and sometimes unrealistic) objectives, inappropriate timing expectations and a desire for political solutions that are hard to come by (Gould & Zobrist 1989: 1717). Actors have also had different views on the purpose of the NBA. Donors have tended to emphasise planning responsibilities, while the riparian states have wanted to put emphasis on project implementation and fund raising (1719).

In a comprehensive study of international river basin organisations in Sub-Saharan Africa, commissioned by the World Bank, the NBA is considered to be a complete failure, much due to lack of interest among member states and too many member states. At the time of the study, there was no activity, even the phones were disconnected due to nonpayment of bills. The study concludes that '[t]he NBA is never likely to succeed in its present form' (Rangeley 1994). Instead, the authors of the report state that it needs to be divided into smaller units, with a direct connection to the needs and interests of the riparians. A later World Bank report emphasise that the NBA lacked all the basic elements of legitimacy, relevance, and constituency, which in their turn impact on capacity and financial viability (Andersen et al. 2005: 61).

Despite such severe critique, the NBA has made a new start. In 2000, the heads of state and government of the riparian countries decided to elaborate a shared vision to fight poverty and promote development in the basin.[9] The main purpose was to 'create an environment conducive to cooperation between Member States and development partners, and to develop an action program accepted by all stakeholders in the basin' (Great Rivers Partnership 2013: 9). The formulation of a shared vision rather than a master plan for resources development is in line with the World Bank emphasis on creating dynamic cooperation by building 'the community of interest and political constituency among and within the member states that is the key to moving cooperative developments forward' (Andersen et al. 2005: 60).

In 2002, the seventh Heads of State Summit, held in Abuja, acknowledged the danger of unilateral planning and committed to pursue a regional dialogue and seek support for cooperative, sustainable development of the Niger River basin (World Bank 2007). In 2004, the heads of state and government signed the Paris Declaration on the principles for management and good governance for a sustainable and shared development of the Niger basin. It stipulates that the member states have to systematically consult each other for every intervention that significantly affects the water regime in the basin, and to strengthen the capacity of the NBA to play its role as coordinator (NBA 2008: 3).

There is a consensus on the need to establish a competent organisation, functional connections between the organisation and the member states, a group of reliable and committed donors, and the participation of civil society. For that purpose, in 2006, a plan to strengthen the capacity of the secretariat was developed and is under implementation with continued support from a range of donors. The same year, the first Niger Basin Civil Society Forum reasserted the needs, aspirations, and dynamics of stakeholders contributing to the cooperative development and management of the Niger basin as a common heritage (World Bank 2007).

The plan of action for sustainable development (SDAP) was signed in 2007, with the purpose of translating the shared vision into activities until 2025.[10] Its prioritised areas are 1) the conservation of the ecosystems in the basin; 2) the development of socio-economic infrastructure; and 3) strengthening the capacity of the stakeholders. The SDAP consists of an agreement between the nine countries on a number of activities, at the centre of which is the construction of three large dams, one in Guinea, one in Mali, and one in Niger. The plan is based on a multi-sectoral assessment of the needs and interests of the nine riparians (NBA 2005). The political process of making actual prioritisations made the negotiations over the SDAP lengthy and complicated, particularly between the two strongest countries in the basin, Mali and Nigeria. New dams will have effects on water flow into Nigeria, thus affecting the functionality of its dams. They also run the risk of rendering obsolete the current electricity contract between Nigeria, Niger, and Benin (Interview project staff NBA 2013).[11] The construction of the Fomi dam in Guinea has also been considered a threat to the Inner Delta of Mali, and important advocacy work has led to the re-design and reduction in scale of the dam (Wetlands International). After environmental assessments and the establishment of minimum flows, all nine riparian states signed the SDAP. However, contentions remain, and during the 2012 floods, Nigeria accused Mali of having released too much water from the Sélingue dam. However, NBA studies showed that the flooding was a result of heavy rains in northern Burkina Faso.

Although there is recognition of the benefits of the SDAP for economic development in the basin, there has been expressed concern regarding the social and environmental costs. Other alternatives – such as small-scale irrigation, solar or wind energy, and ground water use – have been promoted by opponents to large-scale infrastructure development. Hence, there is a need for increasing consultation, and the ECOWAS Water Resources Coordination Centre (WRCC) has launched a consultation initiative in cooperation with the West African Economic and Monetary Union (WAEMU), African Network of Basin Organizations (ANBO), Global Water Partnership-West Africa (GWP-WA), World Wide Fund for Nature (WWF), and the International Union for Conservation of Nature (IUCN), for the West-Africa region (IUCN-PACO 2012).

The Niger Basin Water Charter, adopted in 2008, constitutes the legal framework for cooperation in the basin. Previously, the lack of legal instruments for effective cooperation has, according to the World Bank, been the main reason many countries have focused on unilateral development of the river's resources (World Bank 2007). The Charter stipulates the priorities of the NBA, good governance and sustainable development, as well as solidarity and reciprocity as guiding principles. The Charter gives highest priority to irrigation and domestic water supply to secure food production and alleviate poverty (Grijsen et al. 2013: 50). The agreement between the Niger riparian states is relatively far reaching as it constitutes (at least in principle) an agreement for joint management of internationally shared waters, rather than merely allocating water between states. Cooperation is supposed to be based on basin-wide planning and in the programme for development (SDAP) an assessment of which infrastructure plan is the best for the entire basin provides the basis for decision making.

The Niger basin riparian states, as mentioned above, have been applauded for the commitment they have shown through the achievements of the Paris Declaration, the Shared Vision, and the Water Charter. By in-house donor staff, the NBA is lauded for being an initiative driven by the riparian states rather than coming from outside. Many decision-makers and elected authorities in the basin are also considered committed to reform (Great Rivers Partnership 2013: 17). However, there are many obstacles to the actual implementation of documents, and there are still issues not resolved by the current agreements (10). The Law of the Non-navigational Uses of International Watercourses (UNWC) that governs international watercourses in the absence of applicable agreements has not been ratified by four of the riparian countries. Efforts to fill the gaps in national policies and alignment to the UNWC are carried out through the African Ministerial Conference on Water (AMCOW) and the ECOWAS Water Resources Coordination Center (WRCC).

While the Niger Basin Authority is the main institution for cooperation in the management of the river and its ecosystem, the involved countries have been cooperating for a long time on issues that are closely related to climate change effects, such as desertification, and the challenges of agriculture in semi-arid and arid regions. For example, the Comité Permanent Inter-Etats de Lutte contre la Sécheresse au Sahel (CILSS), since 1974, and its research branch Application en Agrométéorologie et Hydrologie Opérationnelle (AGRHYMET), which includes a project on strengthening the capacities of the member states to adapt to climate change in the Sahel. A collaboration is also under way with ECOWAS member states to develop a West African Climate Change Adaptation Strategy (WACCAS).

Finally, there are five bilateral agreements on the management of the Niger. Niger and Benin have agreed on the development of a hydroelectric power station in Dyongyonga on the Mékrou tributary (1999), and Niger and Mali have agreed on cooperation on the use of the river's water resources (1988). On the Upper Niger, Guinea and Mali are cooperating on a hydroecological management project, and Niger and Nigeria have signed an agreement on equitable sharing, conservation, and development of their common water resources (1990). In 2000, Nigeria and Cameroon signed a Protocol Agreement for the coordination of release of water from dams, and consultation on water structure projects, an agreement facilitated by the NBA (ECOWAS-SWAC/OECD 2006).

To sum up, there is a long history of cooperation between riparian states, but cooperation has suffered from incapacity to respond to the challenges they have been facing. New hope has been placed in the strong initiative and commitment shown recently by both the NBA and riparian countries. Yet, many challenges remain for effective cooperation to take place. The scope of the task to be achieved on several levels and scales increases the risk of institutional fatigue when set targets cannot be met. These challenges become all the more important when dealing with the unpredictability of climate change in the basin, and contribute to shape governance of water resources. In the following, we will see how this problematic is played out in regional and national responses to climate change.

Climate change adaptation at the basin level

Although the documents that govern the activities of the NBA and cooperation in the basin – the SDAP, the Shared Vision, and the Water Charter – are not dominated by a climate change discourse, the climate is certainly at the centre. The measures that address vulnerability in the basin are at the same time considered to build resilience to climate change (Kerres 2010: 24). Meaning that doing what they would have to do anyway is considered to constitute climate change adaptation measures. Moreover, at the eighth Heads of State Summit the challenges of climate change to the goals of the SDAP were raised and NBA requested the World Bank to undertake a climate risk assessment (CRA) 'in an effort to climate-proof new investments' (Grijsen et al. 2013: 52). The assessment is working with a worst-case scenario of 20 per cent reduction in the long-term average basin runoff. Effects on hydroelectric production, navigation, and irrigation is considered limited, while rain-fed agriculture is facing a higher risk. It is also concluded that the construction of dams planned in the SDAP, particularly the Fomi and Kandadji dams, constitute efficient ways to adapt to climate change (Grijsen et al. 2013). There is hesitation as to the benefits of the Taossa dam in Mali in this regard, as it is being constructed in a high-evaporation area (Interviews donor staff NBA 2013).[12]

The three main programmes and projects currently conducted via the NBA and donors do not explicitly define climate change as their main target, but it emerges in project documents as an important motivating force and contributing factor. The programmes and projects include the following.

Le Programme de Lutte contre l'ensablement du fleuve Niger, run by NBA, financed by ADB, and initially covering three countries, Burkina Faso, Niger, and Mali. The programme consists of preventive measures, such as improving rain infiltration in ground water, stabilising sand dunes, and restoration of degraded land. The project is implemented by vulnerable communities in high-risk areas through a participative approach. The first phase has been evaluated and has shown important results in several areas, and the next phase is under negotiation. This programme does not only imply an adaptation to climate change, but an actual mitigation of climate change effects, both as it brings 'lost' land back into production and as it improves the flow of the river.

The World Bank is financing the Niger Basin Water Resources Development and Sustainable Ecosystems Management Project, which has the objective to 'enhance regional coordination, development and sustainability of water resources management in the Niger Basin', with a focus on capacity building, improving performance of existing hydroelectricity plants, of irrigated agriculture, and of watershed management. The project is thereby considered to be contributing to climate change adaptation.

The third large project, Niger-HYCOS, aims to restore and renew observation stations to provide necessary information to estimate developments, to regulate minimum flow, as well as to predict flow levels and prevent and precaution in case of extreme events. In addition GIZ is running an IWRM project that brings institutional support to NBA, and KfW invests in the protection of the Niger River and in sustainable small scale irrigation in Niger.

In the pipeline is a major Program for integrated development and climate change in the Niger River basin, financed by AFDB, UEMOA, KfW, BOAD, LDCF/SCCF, and the NBA countries. It will start in June 2014. Workshops are being held during a pre-phase of the project (Interview project staff NBA 2013).

Several large as well as small programmes and projects are specifically addressing climate change adaptation, but not necessarily with the Niger basin as the limitation of their 'geographical imagination', but focusing on everything from local communities to the Sahel region, or West Africa. Particularly, the CILSS and Agrhymet have been the sub-regional institutions involved in such projects and programmes.

In interviews, the climate change discourse comes through as having affected TWM in a number of ways. First of all, it emerges as a catalyst for intensified cooperation, as it points at the urgency of coming to terms with environmental degradation. A majority of the population within the basin lives on rain-fed agriculture and, hence, is directly affected by changes in rainfall. No rain, nothing to eat. According to an NBA director, the shared vulnerability of the riparian countries have made them realise that there is a convergence of interests to improve the management of shared water resources to face the challenges of long-term decreasing rainfall.

The institutional reform that has been going on for the last 14 years, and the flexibility of the new structures is seen as shaped and informed by climate change, which is also in accordance with current ideas among institutions such as the World Bank as to the need for flexible and dynamic agreements (Andersen et al. 2005). The SDAP is seen by some as a response to climate change effects, which is interlinked with the importance of exploiting the river for the purpose of development. Socio-economic development, it is regarded, is absolutely crucial to overcome the current situation of vulnerability to climate change effects. Moreover, the climate change discourse has effects on the access to financial resources for activities in the basin as it has increased investments in recent years. As one senior water ministry staff puts it:

> All strategies that the state formulate have climate change at the centre. It is related to the decision makers, to the ones who finance. There is a tendency at the World Bank, IMF and some countries that when they have a model everybody has to follow it in order to be funded. So in everything we do they ask that we tie it to climate change.
>
> (Interview senior water official 2013)

One of the directors of the NBA clearly states that climate change effects and the need to exploit the water resource has contributed to make donors willing to finance new dams, but it has also contributed to finding an agreement among the riparian states on the SDAP. He also states that donors have previously hesitated to finance dams when they were conceived on national level because of the risk that they would cause conflicts between countries. However, when conception is on basin level and via the NBA, they have reconsidered. The fact that the projects are conceived by NBA also makes funds available that are not accessible on a bilateral basis. Both through feasibility studies and social and environmental assessments, as well as follow up and control, the NBA is expected to safeguard that dams don't do harm to other riparian countries. Although the primary objective for building dams is still economic and social development, it is evident how the threat of climate change and the protection of ecosystems have a legitimising function, as confirmed in the World Bank/NBA climate change risk assessment (Grijsen et al. 2013). This should not be taken to mean that they are not right, but to show the effects of the environmental debate on what policies are possible to conceive of.

Much focus is also placed on rehabilitation, and making more-efficient use of existing infrastructure, such as the Kainji and Jebba dams in Nigeria, and dams on tributaries in Benin, and to increase efficiency of existing systems for irrigation, such as l'Office du Niger in Mali, for example, as well as on dredging of the river to reduce the effects of siltation on river flow and on functionality of existing dams.

From donors, states, and the NBA, there is thus a strong response to climate change on the regional level. However, there are many obstacles to the implementation of conceived programmes and texts. The major obstacles that are defined by water officials and NBA staff are, on the one hand, the level of implementation of policies and strategies, and, on the other, the sustainability of the NBA. These will be further discussed after we have looked briefly at the national responses to the climate change debate.

National responses

'The Sahelian states recognised the utter importance of combatting drought in their food security and sustainable development policies, way before climate change came to the limelight of international debate' (Kandji et al. 2006: 15). In relation to the severe droughts during the 1970s, the Sahelian states came together to create Comité permanent Inter-Etats de Lutte Contre la Sécheresse dans le Sahel, CILSS, with the purpose to 'invest into the research for food security and in the fight against the effects of drought and desertification in order to achieve a new ecological equilibrium' (CILSS 2004; Kandji et al. 2006: 15). While environmental concerns have shaped national politics for a long time, as mentioned above the centrality of agriculture in development plans during the last 10 years has placed the climate centre stage. All basin riparian states have ratified the United Nations Framework Convention on Climate Change (UNFCCC), and developed their National Adaptation Program of Action to Climate Change (NAPA).[13] Niger was first to produce its NAPA in 2006, and the other followed suit. Burkina Faso, Mali and Guinea in 2007, Benin in 2008, Chad in 2010, and Nigeria, not being categorised as a Least Developed Country, has developed their National Adaptation Strategy and Plan of Action on Climate Change for Nigeria (NASPA), with the support of the Building Nigeria's Response to Climate Change (BNRCC) Project, financed by CIDA, completed in 2011. Ivory Coast has planned for a Climate Change strategy in their Second Communication to the UNFCCC (MEEF 2010). Cameroon has submitted a set of Nationally Appropriate Mitigation Actions (NAMAs) to the UNFCCC in 2010, and is preparing a NAPA. Due to the high number of riparians in the basin, all the NAPAs will not be treated separately here, but they will be discussed in conjunction in terms of how they relate to TWM.

   The NAPAs have the primary objective of identifying urgent activities and raising awareness and developing capacity rather than a focus on implementation (Kerres 2010: 23). The programme sets the principles according to a pre-established format, emphasising participation, multi-disciplinarity, complementarity, sustainability, gender equality, country initiative, and rational management of the environment. As such they identify priority sectors such as agriculture, fisheries, water resources and health, as well as the most vulnerable groups who tend to rely on subsistence production in one or several sectors.

   None of the NAPAs explicitly deal with the Niger River basin, nor do they have a transboundary perspective. Mali, however, emphasise reinforcement of sub-regional cooperation for cross-border water management, as well as establishment of hydroelectricity sites. Mali also has one project concerned with watershed management in the Inner Delta, financed by Wetlands International. Several other countries identify activities that indirectly relate to the river system, such as effective soil and water conservation techniques and micro irrigation. Guinea expresses a concern with decreasing water flows, and focuses on how to better manage water withdrawal through micro dams and pastoral water points.

Niger may be the country that places most emphasis on the water sector in its NAPA. Niger emphasises the need to decrease water demands, as well as to change irrigation practices and improve water management systems and build reservoirs. Niger also prioritises fishing through water control and protection of riversides. In its second communication to the UNFCCC, the imperative of better water management is further strengthened.

Nigeria has as one of its priority adaptation actions to construct more dams and other reservoirs, as well as to desilt reservoirs. Desilting of reservoirs is part of the project World Bank has with NBA, and will improve Nigeria's capacity for producing electricity, both for national and for basin consumption.

An important area for the riparian states to address that will impact on climate adaptation in the region is the strengthening of monitoring and forecasting systems. At the one end, Guinea has no projects addressing lack of climate monitoring and forecasting systems – which can undermine other adaptation efforts (Adaptation Partnership 2011), as well as the work of the NBA in this regard. At the other end, Niger's National Communication on Climate Change, however, places great emphasis on the absence of national technical expertise and information, and aims to focus on building such capacity. Niger also expresses a need for national expertise in managing tools and methodologies for appropriate evaluation of vulnerability and adaptation to climate change. In its UNFCCC communication, Nigeria places much focus on producing and disseminating information through reinforcement of the capacities of the Nigerian Meteorological Agency.

It is important to point out the challenge that the UNFCCC communications have implied in countries that at the time had little experience working with climate change, and that to a large extent lacked the necessary instruments and information required. There is also a severe lack of 'national expertise mastering the tools and methodologies for an appropriate evaluation of vulnerability and adaptation to climate change' (Niger second communication). This has had consequences both on the time needed to develop communications and NAPAs, but obviously also on the quality of the result. It is possible to see in cases where a number of communications and plans have been developed how the quality both of information and of suggested adaptation measures have improved, such as in Niger's second communication, although the list of shortcomings and constraints at the human, material/technical and financial level, that it puts forward, is still extensive.

According to an assessment made in 2011, there is a difference that can be noted in terms of the number of activities in the sector that are registered in the different countries. Guinea and Niger show a lower number of activities, most likely due to political instability and an absence of donor investments. Ivory Coast is also showing fewer activities, which means it is lacking a national strategy and is not being an LDC. Mali has been comparably active when it comes to climate change adaptation in the region in terms of implementation, primarily through bilateral and NGO-supported projects. The review also shows low level of implementation in all countries. All countries are considered weak on capacity to adapt to climate change, in institutional, technical, cultural, and economic terms.

Other plans, such as National Environmental Action Plans (NEAPs), National Action Plans to Combat Desertification (NAPs), the PRSPs, Rural Development Strategies, and Integrated Water Management strategies, also contain elements that deal with climate variability.

Adapt, resist, subvert

So how can we understand responses to GCC in the basin in terms of the analytical framework of this volume; namely in terms of adaptation, subversion and resistance?

Adapt

First of all, the riparian states have shown strong commitment to the reconstitution of the Niger Basin Authority since the early 2000s. In the identification of the World Bank and GEF Niger Basin Water Resources Development and Sustainable Ecosystems Management Project, it is stated that there is a broad recognition among the riparian states that environmental sustainability is linked to 'food production, tourism, sanitation, population movements, and thus regional stability' (World Bank 2007). This realisation has motivated the riparian states to actively participate in development of the NBA, and show commitment to regional efforts to deal with shared environmental concerns. This evaluation is shared and reaffirmed by both NBA and donor staff on the ground. It is further reflected by the adoption of the SDAP that relies on a national multi-sectoral study in each member state to make an inventory of fixtures, to assess the opportunities, constraints, and priorities per development sector in the national portions of the basin (NBA June 2005). It is also reflected in the adoption of the Water Charter.

Rising temperatures and decreasing rainfall in combination with degradation of the environment, as well as a decrease of the flow in the river, makes adaptation measures an absolute matter of survival, both of populations and of states. Adaptation measures are, therefore, not a concern for predictions for the future, but of life and death in the present. It also needs to be taken into account that several countries are in the midst of, or just coming out of, violent conflict that to some extent is resource related, or as we have seen in the above, discussed in terms of resources and climate change. Moreover, the centrality of agriculture in socio-economic development is absolutely interlinked with environmental protection and ecosystem restoration. While construction of dams is high on many countries' wish lists, it is evident that without efforts of desiltation, dredging, restoration of lost soil, investments in such infrastructure will be undone. It remains to be seen whether the construction of the new dams will be accompanied by complementary interventions, which will determine the capacity to mitigate climate change effects.

On the organisation level, there is an attempt to promote resilience and robustness of the institutional framework through the adaptation of the central texts of the NBA. However, in order for NBA to actually achieve the needed resilience and robustness, several issues remain to be solved. The first one is sustainable funding, and the second is a functional organisation. Sustainable funding will be further discussed under the resistance headline.

The other challenge is a functional organisation. Improvements have been made since 2004 and a more or less regular income of contributions from the riparian countries has allowed the NBA to recruit new staff in a number of positions, although far from sufficient. Training is highly prioritised; however, it has been indicated that a continued row of workshops that require staff to leave their positions has serious consequences on the functionality of the institution, as well as on the perceptions that other actors have of the NBA. GIZ, for example, will therefore increasingly focus on on-the-job-training (Interview with GIZ staff 2013).

One organisational audit has been made and another is currently performed to investigate the necessary functions, and levels of decentralisation needed for an effective river basin authority. Sub-commissions are stipulated as a condition in the World Bank project for the second phase to start. However, there are a number of functions performed on different levels and it is of importance not to create double structures but to put in place an efficient organisation (Interview with donors staff 2013).

The institutional framework is developing progressively. For example, as the tasks defined in the SDAP have been considered too vast for the current capacity of the NBA and its member states, a strategic plan was developed in 2010. Moreover, two appendices for the Water Charter are still under discussion, concerning environmental protection and coordinated management of infrastructure. Developing the institutional framework is a political process, and as one donor staff said, it must be allowed to take time.

An additional challenge on the organisational level is the implication of the population. Since the first Niger Basin Civil Society Forum held in 2006, there has been a continued presence of civil society in the ministerial and heads of states meetings. It remains, however, to better include local practices and knowledge in modeling, that up to now have not contained that type of information (Interview World Bank project staff 2013). This is important both for making better predictions but also for knowing what adaptation measures are needed and possible (Interview NBA director 2013). While capacity may be strengthened at the regional level when it comes to dealing with information, early warning mechanisms and disaster response activities, there is often little capacity to respond at the local government level (Goulden & Few 2011). An important, and possibly underinvestigated, issue may be the capacity of local governments to, for example, evacuate areas threatened by flood. This is a concern expressed by local officials in the Nigerian media in relation to the most recent flood in 2012 (Maduforo et al. 2013). There are projects that place the population at the centre though, such as the *programme Lutte contre ensablement*, that rely on the population for implementation and organise regular meetings for exchange of experiences between different groups in the respective countries.

On a systems level, warning systems and dam synchronisation mechanisms are in the pipeline. The former French colonies (all but Nigeria) rely on an old OSTROM network, and there are 105 monitoring stations throughout the basin. In 1984–1965, hydroclimatological data collection platforms were established along the Niger and its tributaries through the HYDRONIGER project. However, the system is not fully functional due to important constraints in terms of financing and management.

Due to the recurrent droughts in the Sahel region since the 1970s, there has been a history of assistance in the domain, however, not primarily on river level. One example is a programme of agro-meteorological assistance in Mali, initiated in 1982, and with the Ministry of Agriculture and the National Directorate of Meteorology (DNM) as host institution. Information about rainfall is collected and disseminated to farmers through radio, TV, and bulletins in local languages. Other monitoring programmes are run by the National Directorate of Water (Direction Nationale de l'Hydraulique, DNH). Few countries have been able to develop such a system though, or they have a system that has been degraded for some time. Constraints are both in terms of equipment and educated personnel (Interview project staff 2013). A large portion of the monitoring stations are situated in Mali and Nigeria, where access has recently been difficult due to violent conflicts.

The flood in 2001 caused severe damages to farmers and residents downstream of the Selingué dam in Mali, and without warning released large amounts of water to save the dam. Farmers sued the management of the dam and won the case, which eventually resulted in a settlement. An additional result was the establishment of a water management committee (CGE) that, in theory, would involve all stakeholders and ensure coordinated management of the dam. Its implementation has been insufficient but there are efforts to improve its functionality. Since the establishment of the committee the disaster of the flooding of 2001 has not been repeated (Goulden & Few 2011: 30).

In 2012, Niger and Nigeria were hard hit by floods that had not been predicted. Nigeria initially accused Mali of having opened its flood gates, however, an assessment of the event conducted by NBA showed that the flood did not come from the Niger mainstem, but from heavy rains in northern Burkina Faso, and an ensuing conflict could be avoided (Interview with donor staff 2013).

On the structural level, we see the SDAP, the construction of three dams on the main stem that makes it possible to regulate water. Regulation of water flow is presented as one of the main justifications for dam construction.

Investments are also made to renew and improve the efficiency of existing dams and irrigation systems, for example the Office de Niger in the Inner Niger Delta. Just as important are the mitigating measures taken through the Lutte contre l'ensablement project and a number of smaller projects that work with reforestation, restoration of land, the development of reduced water-demanding and more-resilient crops, and diversification of agricultural production. Several countries have worked with local dispute associations to deal with increasing conflicts between herders and sedentary farmers. In Nigeria, local government mediators have been put in place to deal with pastoralist/farmer conflicts. For example, grazing channels have been established in order to combine expanding farming with pastoralism. Hard lessons have been learned when it comes to the risk of locking in practices and systems in terms of development activities in ways that do not create the flexibility required to deal with climate change. For example, the droughts in the late 1960s and mid-1970s had particularly severe effects due to the development and modernization of the agriculture undertaken during the relatively wet years in the 50s and 60s. Agriculture expanded into historically marginal lands that would have served as grazing areas for pastoralists in dry years, thus creating systems that were not sustainable in drier years. When the wet years were over, agriculture suffered and conflicts erupted with pastoralists who could no longer access grazing areas (Goulden & Few 2011: 21–22). Moreover, the promotion of private land ownership has favoured agriculture over pastoralism, with conflictual consequences in areas where agriculture and pastoralism, have previously cooperated through reciprocal systems (Lund 2000). Continued desertification in combination with higher temperatures run the risk of increasing southward movements of herders and livestock production, and possible conflicts over access to land.

As we have seen, the Niger basin riparian countries have managed to develop a rather strong legal framework and have given the NBA a regulative role, and actions are planned and implemented on the system and structural level. However, the level of implementation that will safeguard adaptation remains a challenge. In front of the NBA and its member states lies a struggle for prioritisation, funding, and attention of a limited administration. The extent to which resource mobilisation and infrastructure investments, as well as the implementation of agreements between states, will improve adaption to climate change, rather than constituting subversion of, or resistance to, the climate change discourse also remains to be seen.

Resistance

As was argued above, there is little possibility or gain for the Niger riparian states to ignore and, hence, passively resisting the climate change discourse. Planning, allocation, and infrastructure development take climate change into account, both in the shape of adaptation to natural climate variability and with explicit reference to climate change. Nor is climate change questioned, although its main role in increasing vulnerability in the river basin is problematized and other challenges, such as increasing population and anthropogenic activities, are seen as equally important.

The first area where resistance is considered in this volume is with regards to infrastructure investments. Important investments in irrigation are made, not least in the Office du Niger area in Mali, both for sugar cane production through a Sino-Malian joint venture, as well as for small-scale agricultural production throughout the basin. The planned dams also provide important irrigation potential. Although they may threaten the sustainability of the river flow the importance of increased irrigation for agricultural production in a region that faces chronic food shortage emerge as a strong argument and motivation for maintaining existing infrastructure as well as for expanding irrigation. The Office de Niger irrigation system, constructed by the French during colonialism and now mainly used for rice production, withdraws 2.7 cubic metres of water each year, equivalent to 10 per cent of total flow. More recently, the 40 km-long Macina canal has been constructed to irrigate a 100,000 hectare land area that is leased by Libya in the region of Boky Wéré (Bunting 2010).[14] The large land allocation to new investors in the Office du Niger Area might lead to water grabbing, where water is diverted and controlled for the benefit of powerful actors, while increasing the vulnerability of local communities (Hertzog et al. 2012; Kay & Franco 2012). While legitimised in public discourse, these major investments in water withdrawal can be seen as a form of resistance of the climate change discourse.

The second area of resistance concerns incorporation of climate change in international agreements on water management. Here we see how climate change is nominally given a prominent role in the regional agreements, although, as mentioned above, it is presented as one of many destabilising factors that need to be addressed. Moreover, the strategic development plan has been subjected to a climate change risk assessment, hence placing climate change at the centre of cooperation. Being adopted as late as 2008, the Water Charter as the legal framework for cooperation takes into account the importance of flexibility, and taking adequate measures in view of 'the study of the water cycles and inventories per catchment' (NBA 2008). As we will recall, such incorporation has been necessary in order to access funds for water management in the basin. Resistance can rather be seen on the level of implementation of said agreements.

For example, there are indications of lack of political commitment and prioritisation. National water officials as well as NBA, and donor staff laments the relative invisibility of water management issues and the priority given to issues such as oil extraction and mining that more directly fills the national treasury. It is evident that the lack of resources among the riparian states and more-pressing needs may place long-term environmental imperatives far down the priority list (World Bank 2007). Insufficient prioritisation is primarily evident in the lack of funding.

Lack of sustainable funding of the NBA may constitute an obstacle to proposed adaptation measures. The riparian states are currently financing the executive secretariat; however, several countries tend to delay paying their contribution, which causes problems for the functionality of the NBA. Moreover, the secretariat is severely understaffed for its task, and additional funding is required to guarantee a strong institution. For this purpose, in the Water Charter, a polluter-pays principle, as well as user-pays principle are established. However, the riparian states have not yet fully agreed to hand over the tax prerogative to NBA. Moreover, there is very limited access to functioning laboratories in the basin, which is a requirement for the polluter-pays principle to be working. There is also inadequate funding for monitoring equipment, which is expensive and, as it is an activity that is not very visible, it may be difficult to mobilise funds (Interview with project staff NBA 2013).

The IWRM project funded by GIZ is also supporting the effort to adapt and update national texts so that they harmonize with regional agreements. However, the process of doing so is slow. Although this is explained by the lengthy process that involves decision making on many levels, it may also be interpreted as a form of resistance to sufficiently prioritise climate change adaptation.

Subversion

The centrality of climate change in global discourse makes it impossible to access major funds without integrating climate change adaptation in national and regional policies and strategies. Hence, it is possible to interpret current emphasis on climate change in documents and agreements, as well as in motivations for infrastructure projects, as a form of subversion of the discourse. Subversion, in this volume, is taken to mean that the initial meaning of the discussion is transformed in order to benefit certain actors. What we see in the above is that states as well as the NBA have responded by taking initiative for increased cooperation, and inclusion of the imperative to meet the challenges of climate change, and by doing what is asked of them in terms of policy and institutional reform.

Climate change is, for example, put forward as an urgent motive for regulating the flow of the river by construction of new dams. This can be interpreted as subversion in the case of all the proposed dams, as the climate change discourse is used by certain actors to pursue activities for other purposes. Moreover, while the climate change impact assessment supports two of the dams, in Guinea and Niger, as essential in this regard, one of the dams, the Taoussa dam in Mali, may not be justified for this purpose. The climate change discourse can thereby be seen as a way for the Malian government (as well as for the Nigerien and the Guinean) to access funds for other purposes by associating it with the need to regulate the flow of the river as a whole. In addition, the possibility that Mali will try to expand the scope of the Taoussa dam, and thereby jeopardize minimum flow, could be interpreted as a subversion of the climate change discourse (Interview with donor staff 2013).

Conclusion

The climate change debate is playing a significant role in TWM in the Niger basin. However, we have seen that its importance is shaped by the fact that climate variability has been a general characteristic in the basin over time, by the uncertainty of climate change predictions, and the understanding that anthropogenic activities and development have equally important impact on the ecological system of the basin. Nevertheless, the strong focus on agriculture as an engine for development and reduction of poverty and vulnerability make climate change effects an urgent concern for all riparian countries, as well as for external actors that support institutional reform for improved transboundary management of the basin.

As indicated, certain variables contribute to the specific way in which the climate change debate shapes management of the Niger River basin. The first aspect concerns the long-term variability and the fact that it has motivated cooperation in the Sahel region since the 1960s and 1970s. There is thereby a strong awareness of the importance of dealing with climate variability both on a national and regional level. On the one hand, there are existing mechanisms and measures of adaptation, although not necessarily well implemented; on the other hand, climate change is conceived of as adding to an already-harsh and vulnerable situation. The second aspect concerns high levels of poverty and vulnerability of the populations in the basin. This means that large-scale infrastructure investments are broadly considered legitimate in order to improve agricultural and energy production. The rather limited exploitation of the river system also contributes to reduce opposition against proposed infrastructure. A strong reliance on agricultural activity in poverty-reduction strategies on the one hand makes it impossible to ignore the effects of climate change, and on the other hand further legitimises infrastructure investments. Moreover, there is an important risk that levels of poverty contributes to place long-term environmental sustainability far down the list of priorities. The third aspect that shapes water management concerns weak implementing capacity, which significantly reduces the capacity of the riparian states to adapt to climate change.

All of these aspects are relevant for an analysis of national and regional responses to global climate change in terms of adaptation, resistance, and subversion. What we see is that the climate change debate has contributed to reinvigorating cooperation and strengthening the Niger Basin Authority, as well as to instigating activities on the organisational, systems, and structural levels. However, lack of resources and capacity, as well as lack of prioritisation in face of poverty constitute constraining factors, which can be interpreted in terms of resistance. Finally, we see that climate change provides an opportunity for subversion of the debate to legitimise and finance measures such as dam construction, which may be motivated by other interests.

Notes

1 A small part of the basin, although not active, is on Algerian territory.
2 50 km^2 in Sierra Leone.
3 The Bani has its source in Ivory Coast and passes through Burkina Faso.
4 Just as other humid areas in West Africa, the inner delta has severely diminished since the 1960s. The inner delta has decreased from 37,000 km^2, to 15,000 km^2 today. Not only lack of rainfall underlies the decrease, but also the construction of dams upstream, more precisely the dams at Sélingé and Markala (Barbier et al. 2009: 37).
5 The Benue has its source in Chad but gets most of its flow from tributaries in Cameroon.

6 Figures vary, other sources estimate a population of 94 million in 2005, and a growth rate of 3.2 per cent UNDP in the *2002 Human Development Report* estimates a population of 150 million in 2015. Variations are a result of how fertility rates develop (Ogilvie et al. 2010). According to the Niger Basin Authority (NBA), the population in the basin is estimated to have increased to 190 million by 2027, which constitutes the end of the investment plan of the NBA (NBA 2008: 11).

7 This inability to project climate change in West Africa is explained by different factors, 'parameterization and untested assumptions of the models, inherent variability of the earth's climate system and uncertainty of future emissions', as well as methodical errors in representing the climate in Africa, for example by the misplacement of the Atlantic Inter-Tropical Convergence Zone (ITCZ), or West African Monsoon, as well as under-estimation of climate feedbacks by vegetation (Ghile et al. 2014).

8 Algeria has since then received observer status in the NBA.

9 With support from the World Bank, the African Development Bank, France, Canada, and more recently the EU.

10 The SDAP was realised in two phases, the first one consisting of a diagnosis of the situation, financed by Agence Français de Développement (AFD), and the elaboration of the master plan, financed by the EU.

11 Niger and Nigeria, as well as Benin and Nigeria, have had bilateral agreements since 1972 that regulate the export from Nigeria of electricity to Niger and Benin, agreements that have been re-negotiated and extended and are still in force (World Bank 2007 Report No 43582). The NBA investment plan in new dams will increase energy security in Guinea, Mali and Niger, while in the short run it will reduce production (and, hence, export) in Nigeria due to limitation on flow. New dams in Nigeria (Mambila and Zungera) will, however, increase production in the long term (NBA 2008: 11). ECOWAS and Canada and UN (UNIDO) have developed a strategy for electric power interconnections in West Africa, the West African Power Pool (WAPP-APL 3, CR 4092 GH- Nov. 1, 2005), which aims to avoid subjecting management of the resource to conflicts over electricity production (Gould & Zobrist 1989: 1720).

12 The Taoussa dam can also be questioned in regards to economic benefits. The most cost efficient is to build dams in the humid areas of the river; electricity produced there could be sold to the Sahelian countries at a lower price than it would cost to produce it there (Barbier 2009: 37).

13 National Adaptation Programs of Action, introduced at the 2001 UNFCCC conference. Application for least-developed countries whereby they 'identify priority activities that respond to their urgent and immediate needs to adapt to climate change' (MEEF 2010).

14 'The Republic of Mali is committed to deliver un-restricted licenses to the project to use water form the Boky-Wéré branch to meet the project's water demand' in the Convention d'investissement dans le Domaine agricole entre La République du Mali et La Grande Jamahiriya Arabe Libyenne Populaire et Socialiste, May 2008 (Hertzog et al. 2012: 314).

References

Abrate, Tommaso, Hubert, Pierre, and Sighomnou, Daniel, 2013, 'A study on hydrological series of the Niger River', *Hydrological Sciences Journal*, 58(2): 271–279.

Adaptation Partnership, 2011, *Review of Current and Planned Adaptation Action: West Africa*. Adaptation Partnership, International Institute for Sustainable Development.

Andersen, I., Dione, O., Jarosewich-Holder, M., and Olivry, J-C., 2005, *The Niger River Basin. A Vision for Sustainable Management*. Washington DC: The World Bank.

Barbier, Bruno, Yacouba, Hamma, Maiga, Amadou Hama, Mahé, Gil, and Paturel, Jean-Emmanuel, 2009, 'Le retour des grands investissements hydrauliques en Afrique de l'Ouest: les perspectives et les enjeux', *Géocarrefour*, 84 (1–2): 31–41.

Bunting, M., 2010, 'Mali: whose land is it anyway?' *The Guardian,* 28 December 2010, accessed 16 February 2015 at http://www.theguardian.com/world/2010/dec/29/mali-farmers

CILSS, 2004, www.cilssnet.org/anglais/index_anglais.htm

ECOWAS-SWAC/OECD, 2006, 'Atlas on regional integration in West Africa: transboundary river basins', accessed 16 February 2015 at http://www.oecd.org/swac/publications/38409569.pdf

Ghile, Y. B., Taner M. Ü., Brown, C., Grijsen, J. G., and Talbi, Amal, 2014, 'Bottom-up climate risk assessment of infrastructure investment in the Niger River basin', *Climatic Change,* 122: 97–110.

Giannini, A., Saravanan, R. and Chang, P. 2003, 'Oceanic forcing of Sahel rainfall on interannual to interdecadal time scales', *Science,* 302: 1027–1030.

Gould, Michael S., and Zobrist, Frederick A., 1989, 'An overview of water resources planning in West Africa', *World Development,* 17(11): 1717–1722.

Goulden, M., and Few, R., 2011, *Climate Change, Water and Conflict in the Niger River Basin.* USAID, International Alert and University of East Anglia.

Greaf, F., and Haigis, J., 2001, 'Spatial and temporal rainfall variability in the Sahel and its effects on farmers' management strategies', *Journal of Arid Environments,* 48: 221–231.

Great Rivers Partnership, 2013, Great Rivers Partnership, Phase II. Volume II Full Basin Profiles.

Grijsen, J. G., Tarhule, A., Brown, C., Ghile, Y. B., Taner, Ü., Talbi-Jordan, A. . . . Harshadeep, B. N., 2013, 'Climate risk assessment for water resources development in the Niger River basin part I: context and climate projections', in Tarhule, A. (ed.), *Climate Variability – Regional and Thematic Patterns.* Intech Open Access.

Hertzog, T., Adamczewski, A., Molle, F., Poussin, J.-C., and Jamin, J-Y, 2012, 'Ostrich-like strategies in Sahelian Sands? Land and water grabbing in the Office du Niger, Mali', *Water Alternatives,* 5(2): 304–321.

IUCN-PACO, 2012, *Regional dialogue on large water infrastructure in West Africa: Building multi-stakeholder participation from 2009–2011,* accessed 16 February 2015 at http://cmsdata.iucn.org/downloads/version_anglaise.pdf

Kandji, A. T., Verchot, L., and Mackensen, J., 2006, 'Climate change and variability in the Sahel Region: Impacts and adaptations strategies in the agricultural sector', United Nations Environment Program and World Agroforestry Centre (UNEP and ICRAF).

Kay, S., and Franco, J., 2012, *The Global Water Grab: A Primer.* Amsterdam: Transnational Institute (TNI).

Kerres, Martin, 2010, 'Adaptation to climate change in the upper and middle Niger River basin: River basin snapshot'. Report commissioned by KfW Entwicklungsbank.

Lund, C., 2000, 'African land tenure: questioning basic assumptions'. Dryland Issue Paper no. E100. London: International Institute for Environment and Development (IIED), accessed 16 February 2015 at www.iied.org/pubs/pdfs/9023IIED.pdf

Mahé, Gil, and Paturel, Jean-Emmanuel, 2009, '1896–2006 Sahelian annual rainfall variability and runoff increase of Sahelian Rivers', *Comptes Rendus Geosciences,* 341(7): 538–546.

Mamadou, D., 2001, 'L'état et la coopération internationale: de quelques aspects de la coopération bilatérale du Niger', in Idrissa, Kimba (ed.), *Le Niger: État et démocratie.* Paris: L'Harmattan.

Ministry of Environment, Water and Forest (MEEF), 2010, *Second National Communication to the UNFCCC.* Yamoussoukro: MEEF.

Mounir, Z. M., Chuan, M. M., and Amadou, I., 2011, 'Application of water evaluation and planning (WEAP): a model to assess future water demands in the Niger River (in Niger Republic)', *Modern Applied Science,* 5(1): 38–49.

NBA, 2005, 'Regional synthesis report on national multi-sectoral studies', *Final Report,* Niger Basin Authority, June 2005.

NBA, 2008, 'La Charte de l'Eau du Bassin du Niger', Niger Basin Authority.

Obi, C. I., 2008, 'Nigeria's foreign policy and transnational security challenges in West Africa', *Journal of Contemporary African Studies*, 26(2): 183–196.

Ogilvie, Adrew, et al., 2010, 'Water, agriculture and poverty in the Niger River basin', *Water International,* 35: 5.

Olsson, L., Eklundh, L., and Adrö, J., 2005, 'A recent greening of the Sahel – trends, patterns and potential causes', *Journal of Arid Environments,* 63: 556–566.

Olukoshi, A. O., 2001, *West Africa's Political Economy in the Next Millennium: Retrospect and Prospect.* Dakar: CODESRIA.

Okpara, Juddy N., Tarhule, Aondover A., and Perumal, Muthiah, 2013, 'Study of climate change in Niger River basin, West Africa: Reality not a myth', in Singh, Bharat Raj (ed.), *Climate Change – Realities, Impacts over Ice Cap, Sea Level and Risks.* InTech Open Access.

Paeth, H., Fink, A., and Samimi, C., 2009, 'The 2007 flood in sub-Saharan Africa: spatio-temporal characteristics, potential causes, and future perspective', EMS Annual Meeting Abstracts, 6: EMS 2009–2103.

Rangeley, R., Thiam, B., Andersen, R., and Lyle, C., 1994, 'International river basin organizations in Sub-Saharan Africa', Washington DC: World Bank Technical Paper, No. 250: 43–48.

Refugees International (RI), 2013, 'Sahel: recurrent climate chocks propel migration; resilience efforts face challenges', Field Report, 1 August 2013.

RFI, 27 April 2004, 'La bonne gouvernance sauvera le Niger', accessed 16 February at www.rfi.fr/actufr/articles/052/article_27673.asp

Rönnbäck, A-S., 2008, 'ECOWAS and West Africa's future: Problems or possibilities?', Umeå Working Papers in Political Science, 3.

Shackley, S., and Wynne, B., 1996, 'Representing uncertainty in global climate change science and policy: boundary-ordering devices and authority', *Science, Technology & Human Values,* 21(3), 275–302.

Taylor, I., and Williams, P. D., 2008, 'Political culture, state elites and regional security in West Africa', *Journal of Contemporary African Studies,* 26(2): 137–149.

United Nations Development Program (UNDP), 2011, 'Fast facts: UNDP and climate change in Africa'.

Wetlands International, 'The Niger River and its deltas', accessed 16 February at http://africa.wetlands.org/Whatwedo/NigerRiverandDeltas/tabid/2944/language/en-GB/Default.aspx

World Bank, 2007, 'Project information document, appraisal stage, Niger basin water resources development and sustainable ecosystems management project', Report No. 43582.

News

Elhadj, Ibrahim, September 2012, 'Aides en faveur des populations: L'opposition dans les panneaux', *La Roue de l'Histoire,* 628.

Enahoro, Eugene, 15 October 2013, 'The globalization of indifference', *Daily Trust.*

Maduforo, Okey, Oweh, I., Nnadozie, Chinwendu, Moses, Dele, and Inyang, Bassey, 27 September 2013, 'Panic in Onitsha, Asaba, others as River Niger overflows bank', *Daily Independent.*

Souleymane, Laouali, 6 September 2013, 'Visite du Président de l'Assamblée Nationale sur les sites inondés de Saga, Liboré et N'Dounga: Soutien et réconfort aux populations sinistrées', *Le Sahel.*

Tope, Ogunde Shola, 14 April 2013, 'Changing rainfall patterns in Nigeria and global warming', *Vanguard.*

6   The Nile River basin

Introduction

Background

The Nile basin is a complex hydrological and hydropolitical system. In terms of hydrology, the Nile is characterised by several tributaries crossing different climate zones, uneven patterns of rainfall (intra- and inter-annual variability) and evapotranspiration rates, and very prone to drought and floods events. The Nile basin has always been highly vulnerable to impacts of climate variability, as river flows are very sensitive to small changes in the average rainfall in the upstream catchments, in particularly in the Eastern Nile sub-basins. In hydropolitical terms, complexity is paramount: a river basin shared by 11 countries, with a long history of political/diplomatic conflicts, absence of comprehensive legal and institutional arrangements, power asymmetry between downstream and upstream riparian states, and a long-lasting situation of hydro-hegemony. In hydro-economic terms, most of the Nile riparians are still very much agriculture-based political economies, with an extreme dependence on water/rainfall availability, with fast-growing needs in terms of food and energy security due to its rapid population an economic growth and societies' changing consumption patterns.

Adding to all these elements of complexity, the region has been experiencing significant changes in economic and political terms for the past five years – a new riparian state (Republic of South Sudan), social and political upheavals in Egypt, the discover of new mineral resources (e.g. oil in Uganda, gas in Rwanda), new investment dynamics (e.g. infrastructure development, foreign direct investment in large-scale agriculture), and new hydraulic infrastructures (such as large-scale hydropower dams in Ethiopia, Sudan, and Uganda). These changes are already having impacts in the power of balance between the Nile riparians and are expected to become even more significant in the years to come. This chapter aims at understanding how is the climate change discourse being factored in the current hydropolitical transboundary configurations in the Nile basin, namely as a key (discursive) element in the cooperation process.

Politicisation of climate change in the Nile basin

The Nile basin is a very politically sensitive transboundary river basin in which several issues (hydrological information, water allocations, legal principles and rights, water and energy demands, development of infrastructures, etc.) are subject to politicisation and securitisation processes, i.e. processes in which governments transform water issues into a matter of top politics and even national security agenda (Zeitoun & Warner 2006). Downstream and upstream riparians alike, in their national and regional hydropolitical relations, tend to make certain issues jump the queue of political priority often based on vested interests – either to maintain a *status quo* situation or to challenge it. Egypt, for example, has often elevated the issue of 'water scarcity' to a national security issue, partially to justify its current water use and to protect existing legal agreements (Cascão & Zeitoun 2010). Similar examples of politicised discourse can be found upstream, whereas riparians establish a direct correlation between increasing utilisation of the Nile water resources and national socio-economic growth development (Nicol & Cascão 2011). These sanctioned discourses have been an extremely influential factor on the position of the riparian countries vis-à-vis their neighbours, and to a greater extent it also has influenced the expectations of downstream and upstream riparians towards the multilateral cooperation process (Earle et al. 2013). Having in mind the abovementioned main storylines, the climate change debate/discourse, essentially because of the uncertainty and ambiguity associated with it, it has not fit immediately the sanctioned discourse of any of the Nile riparian states, be it upstream or downstream. Indeed, the scientific 'uncertainty' on future water availability in the Nile basin, the most critical piece of information in this politically charged basin (see next sections), have allowed Nile countries to use the climate change discourse in a fuzzy manner, and mainly to reinforce existing hydropolitical narratives.

   The available models for the Nile basin put in evidence very different trends for the projected changes in precipitation, which has allowed the countries to emphasise or de-emphasise certain aspects related to future climate change. Therefore, climate change discourse became part of existing water politicisation processes, by using tactics of emphasising or de-emphasising certain information (Zeitoun & Warner 2006). For example, Egypt prefers to emphasise the fact that projections of decreased precipitation can contribute to intensify water scarcity in the downstream country (see Abu-Zeid & Abdel-Dayem 1992; Elwan et al. 2002; MWRI 2005), an argument that contributes to reinforce Egypt's sanctioned discourse. A recent official document released by Egypt's Ministry of Foreign Affairs clearly shows how climate change is being politically used as an argument to reinforce the sanctioned Egypt's water-security nexus. But at the same time, Egypt de-emphasises the importance of developing storage upstream to regulate water availability and anticipate coping strategies to deal with future water shortages (which is part of upstreamers' discourse), even when independent studies show that this could be the right strategy to address future climate change at the basin level (Jeuland 2010).

In brief, all Nile riparian states agree that climate change needs to feature more prominently in the national water policies, but the same cannot be said about their position towards tackling the challenging climate changes issues at basin level, instead of or additionally to the national-based efforts. However, as this chapter will show, other non-state actors have been very keen to bring climate change (namely specific studies, adaptation strategies, funding, etc.) to the forefront of TWM in the Nile basin, namely the Nile Basin Initiative (NBI) institutions and the external development partners that have been providing technical and financial assistance to the multilateral cooperation process for the past decade. One of the main goals of this chapter is exactly to understand how this new transboundary 'approach' to climate change has been subject to adaptation, resistance, and subversion.

Institutional arrangements

Despite the lack of a comprehensive and permanent legal and institutional framework in the Nile basin, there are already institutional arrangements to deal with TWM in the region – namely the NBI and its subsidiary institutions (ENTRO and NELSAP-CU, in the Eastern Nile and Equatorial Nile sub-basins, respectively) and the Lake Victoria Basin Commission (LBVC). These inter-governmental institutions have mandates to deal with transboundary issues. In the case of LVBC, this is specialised institution of the East Africa Community (EAC) and its mandate derives from a legal agreement between the five EAC countries (Kenya, Uganda, Tanzania, Rwanda, and Burundi) signed in 2003 (LVBC 2003). The NBI was established in 1999 as a transitional arrangement, to pave the way for the establishment of a permanent river basin organisation (RBO), once a legal and institutional framework would be adopted by the riparians (NBI 2002). The transition to a permanent river basin commission is still pending the ratification of the legal and institutional framework (Cooperative Framework Agreement, CFA), a process that has been marred by endless political and diplomatic disputes since 2007. This chapter does not aim to analyse the process itself, rather it will only look briefly at its links to the climate change discourse.

Both the LBVC and the NBI include several programmes and investment projects aimed at improving water management at the basin, regional, and transboundary level, that includes not only water management per se, but as well water-related sectors such as agriculture, energy, environment, fisheries, etc. Climate issues have been as well incorporated in the programmes of these transboundary institutions, as it will be analysed further in this chapter. In general, the perceptions that future climate change might change the hydrological regimes, and ultimately even affect the hydropolitical regime, of the Nile and its tributaries and that in result climate change could negatively impact livelihoods at national and regional levels, is being internalised by the transboundary institutions. But how can this process of internationalisation, and its impacts at policy level, be measured? Do programmes and projects of these institutions address the issue of climate change as a cross-cutting problem, or are they mainly minor components of existing projects? Do they envisage the adoption of policies and frameworks at national and regional levels to deal with climate change in a holistic way? Do agreements signed between countries also include provisions on climate change and its possible implications on the water availability, water allocation, operation of hydraulic infrastructures, and transboundary cooperation at large?

Taking into account the level of uncertainty provided by the climate change models available, it is likely that politicisation processes also occur in the processes of negotiating and establishing the transboundary arrangements (institutions, agreements, policies) on the ground. Nonetheless, institutions like the NBI are adopting a 'no-regrets' type of strategy, i.e. a strategy in response to the threats of climate change in order to reduce the impacts of global warming, sea level rise, precipitation changes, etc., that even if the threat of climate change is not as pronounced as it is now feared, there is no need to be any regrets because the basin, the countries, the river will benefit from those measures in any case. Basically a 'no-regrets' approach' includes activities that yield benefits even in the absence of climate change (Heltberg et al. 2009).

GCC in the Nile basin

Climate variability – an old issue in the Nile basin

The topic of climate variability and change is not an unfamiliar issue in the Nile basin. Climate variability has always been in the agenda and discourse in the region – as variability is the pattern (and not the exception) in the basin for immemorial times. Most likely, the Nile basin has the oldest and longest record of studies on the river flows, which was initiated in the pharaonic periods in Egypt exactly to understand the flow patterns and its implications to agricultural production; and centuries later had been replaced by modernised systematic hydrological measurement systems and infrastructures during the Ottoman and British presence in the region (Tvedt 2000). These studies have allowed Egypt, the most downstream and dependent riparian in the basin, to gather systematic information about the uneven quantity of water reaching to its territory and to develop mechanisms to cope with the annual variation (high flows during three months of the year, and low flows in the remaining months), and the inter-annual variability. Extreme climate events, such as floods and droughts, that have occurred in the last two centuries are well documented and to a certain extent have informed the water policies, in particular of the two downstream riparians.

More recently, and to a greater extent, 'emblematic events' such as the droughts of the 1970s and 1980s and the floods in the 1990s have been instrumental in policy change. Examples of it can be found in terms of hydraulic infrastructure planning, for example with the construction of dams and reservoirs in Egypt and Sudan during the 20th century, such as the

Aswan, Jebel Aulia, and High Aswan dams, in order to cope with the issue of 'timely water' and minimise the uneven availability of water for agriculture throughout the year (Chesworth 1994). Other examples of extreme climate events influencing policy-making consist of social engineering, which major example is the controversial resettlement policies in Ethiopia with the goal of moving people from drought-prone highlands region to the so-called fertile 'virgin' lowland areas, in the aftermath of the droughts of the 1980s (Pankhurst 1992; de Waal 1997). But extreme events also took the shape of institutional change, for example with the establishment of the Intergovernmental Author-ity on Drought and Development (IGADD) in the mid-1980s. IGADD was established in a critical period when several Horn of Africa countries were fac-ing severe climate-related consequences, such as ecological degradation, food insecurity and negative impacts on livelihoods – IGADD aimed at providing a regional approach to supplement national efforts. In brief, climate variability has been for long an influential factor, or even driver, of the policies of Nile countries and regional institutions.

The 'new' global climate change discourse

When Global Climate Change (GCC) discourse gained momentum worldwide in the early 1990s, the Nile policy-makers were not indifferent to it. This has coincided with a shifting political context in the region, marked by rapid popu-lation growth, migrations from rural to urban areas, changes in land use, increas-ing demands for energy, and growing pressures on the national governments to promote socio-economic development and distribute the benefits more equita-bly (Howell & Allan 1994). But the emergent GCC discourse has also coincided with an increasing awareness of the climate-induced threats in the Nile basin as extreme climatic events in the region were becoming more persistent and hav-ing severe impacts on the livelihoods of the populations (Hulme 1994; Conway 1996). Indeed, droughts, and floods affect most of the Nile riparians almost every single year, although the impacts are not felt in the same extension and manner by all the countries upstream, midstream, and downstream, as some have more and better early-warning systems, adaptation policies, and infrastructures in place than others.

As it will be analysed in next sections of this chapter, the GCC discourse has been 'adopted' by the Nile riparian states in two different and divergent manners: 1) to justify the governments' lack of capacity to deliver water-related socio-economic benefits, having climate change turned into a perfect scapegoat; and 2) as an opportunity and justification for the urgent need to construct large-scale infrastructures (including water reservoirs) in order to adapt and mitigate the GCC impacts. The perception that climate change is a cross-border issue and that climate change adaptation measures will be more effec-tive when undertaken in coordination (joint measures) with the neighbouring countries is a more recent trend, but that is becoming a central feature of the transboundary institutions and programmes (discourse and action), as analysed in detail in section 3.

Discourses over climate in the Nile basin

As in other regions of the world, climate change in the Nile basin is a hot and controversial topic. Scientific and popular views on the topic do not always coincide, pessimists and optimists clash over projections and forecasts, governments tend to instrumentalise climate change studies for political gains, external partners (such as development agencies) have been putting a lot of efforts for the adoption of climate change mitigation and adaptation plans at national level, and lately national and regional institutions have start looking at it as a great opportunity to mobilise financial resources for programmes and infrastructures. This section looks in more detail at the national level, and section 3 looks specifically at how GCC discourse is influencing the discourse and action of the transboundary water institutions in the Nile region, in particular the NBI and its subsidiary institutions.

Science and media

There are six climate models about the future impacts of climate change and its impacts in the precipitation and water availability in the Nile basin by the 2050s. The conclusions of the scientific studies exhibit a great disparity in terms of projections – while a warming trend is common in all models, some predict a drier and others a wetter climate, although the majority shows that it is more likely that precipitation will increase in the upstream catchments of the Nile tributaries (NBI 2012b). But all in all, all the six studies highlight that regardless what will be the scenario there is a strong need for the countries to move forward with plans to adapt and mitigate the impacts of climate change – in particular because climate variability is a constant in the region and, as such, any action (regardless the actual climate change impacts) can be considered as 'no-regrets' action. On the other hand, the attention given by the media to climate change issues has been growing since the 1990s, but not always necessarily based on appropriate scientific information and often ignoring the fact that climate variability is not something new in the Nile basin. One of the interesting tendencies of the media in the Nile basin is to give more coverage to the droughts than to the floods, although both climatic events are common and dramatic in the region. This is in contrast to one of the Orange-Senqu case (see Chapter 7) where flood events tend to attract most media attention. The focus on droughts is in line with the common sanctioned discourse that the region is (and will be) experiencing a situation of water scarcity – although the scientific basis of the assumptions are seldom debated by the media. In general, the tone set by the media is used and useful for politicians to deny responsibility for the social and economic impacts of the climatic events. It is easier to blame climate change than to assume responsibility for not having the right mechanisms in place to deal with the events.

Governments

The response at the national level from the governments of the Nile riparian states is diverse, according to the political priority given to the topic of climate change. Most of the Nile riparian states have already adopted National Adaptation Programme of Action (NAPAs) in their national policies in order to promote climate change adaptation in their countries. NAPAs are only for LDC countries, which means that Egypt and Kenya do not have NAPAs but they do have similar institutional frameworks to deal with climate change issues. Kenya and Egypt have detailed similar aims in their national climate change response strategy national communications respectively. Table 6.1 shows the project profiles of the NAPAs of the Nile countries, and it highlights the projects directly related to water (although not necessarily with the Nile water resources).

It is interesting as well to observe what are the vulnerable sectors prioritised in the NAPAs by the Nile riparian countries, as it indicates that water resources is common to all of them (Table 6.2), and that for most of them water resources management is one of the priority adaptation projects. Several of the other sectors are vulnerable by virtue of changes to rainfall and runoff (UNEP 2010)

External actors

The adoption of the Climate Change National Plans and Action Plans in the Nile countries is a process that has been to a great extent influenced by external actors, such as UN institutions, working directly on climate topics, such as UNFCC, UNEP, UNDP, but as well institutions such as the Global Environmental Facility (GEF), a World Bank specialised institution. Global lobbying and advocacy campaigns had been an essential leverage for the adoption of the NAPAs, among other climate change adaptation mechanisms. UNDP Africa Adaptation Programme, a US$92.1 million programme designed to assist 20 countries across Africa (including Ethiopia, Kenya, Tanzania, and Rwanda) to incorporate climate change risks and opportunities into their national development processes in order to protect development gains from climate change. But one of the best-known multilateral funding streams for adaptation is the Least Developed Country Fund (LDCF) developed by UNFCC (all Nile countries have benefitted, except Kenya and Egypt), which financial mechanism has been managed by GEF (Prowse & Snilstveit 2010). As of September 2012, the Fund had approved $353.66 million for the funding of 78 projects and programs, leveraging $1.65 billion in co-financing. Simultaneously, there has been a dramatic increasing in the financial support from bilateral development agencies (through financial mechanisms like the UNFCCC, but as well through existing channels of development assistance) to the inclusion of climate change in national policy discourse and practice, including in water-related projects (Agrawala & Aalst 2008). In the particular case of the Nile basin, it is possible to observe that in result of donors' redefinition of aid/development agendas and later the political deadlocks in the Nile cooperation, funding for climate change

Table 6.1 Priority projects of NAPAs of Nile riparian states

Country	Programme profile	Year of adoption
Ethiopia		2007
	• Promoting drought/crop insurance	
	• Strengthening/enhancing drought and flood early warning systems	
	• Development of small scale irrigation and *water* harvesting schemes in arid, semi-arid, and dry sub-humid areas	
	• Improving/enhancing rangeland resource management practices in the pastoral areas	
	• Community-based sustainable utilisation and management of wet lands	
	• Capacity building program for climate change adaptation	
	• Realising food security through multi-purpose large-scale *water* development project in Genale–Dawa basin	
	• Community Based Carbon Sequestration Project in the Rift Valley System	
	• Establishment of national research and development (R&D) center for climate change	
	• Strengthening malaria containment program (MCP)	
	• Promotion of on farm and homestead forestry and agro-forestry practices in arid, semi-arid and dry-sub humid parts	
Burundi		2007
	• Improvement of seasonal early warning climate forecasts	
	• Rehabilitation of degraded areas	
	• Safeguarding of the natural environments	
	• *Rainwater* Valorisation	
	• Erosion control in the region of Munirwa	
	• Protection of buffer zones in Lake Tanganyika floodplain and around the lakes of Bugesera	
	• Popularisation of short cycle and drought-resistant food crops	
	• Zero grazing technique	
	• Capacity building to promote energy-wood saving techniques	

Country	Details	Year
Rwanda	• Stabilisation of river dynamics of ***watercourses*** and torrents in Mumirwa, including the city of Bujumbura • Education on climate change adaptation • Increase hydropower micro stations • Lands conservation and protection against erosion and floods at districts level of vulnerable regions to climate change • Mastering hydro meteorological information and early warning systems to control extreme phenomena due to climate change: Installation and rehabilitation of hydrological and meteorological stations • Development of irrigated areas by gravity ***water*** systems from perennial streams and rivers in zones often vulnerable to prolonged droughts • Support to districts of vulnerable regions to climate change in planning and implementing measures and techniques related to land conservation, ***water*** harvesting and intensive agriculture, and promoting existing and new resistant varieties of crops adapted to different bioclimatic soil • Increase adaptive capacity of grouped settlement 'Imidugudu' located in vulnerable regions to climate change by the improvement of potable ***water***, sanitation, and alternative energy services, and the promotion of nonagricultural jobs • Increase food and medicine modes of distribution to respond to extreme climate change and sensitise to stocking and conservation of agriculture products • Preparation and implementation of woody combustible substitution national strategy to combat deforestation and erosion as well	2006
Tanzania	• Improving food security in drought-prone areas by promoting drought-tolerant crops • Improving ***water*** availability to drought-stricken communities in the central part of the country • Shifting of shallow ***water*** wells affected by inundation on the coastal regions of Tanzania mainland and Zanzibar • Climate change adaptation through participatory reforestation in Kilimanjaro mountain • Community-based mini-hydro for economic diversification as a result of climate change in same district • Combating malaria epidemic in newly mosquito-infested areas	2007
Sudan	• Enhancing resilience to increasing rainfall variability through rangeland rehabilitation and ***water*** harvesting in the *Butana* area of *Gedarif* State • Reducing the vulnerability of communities in drought-prone areas of southern Darfur State through improved ***water*** harvesting practices	2007

(Continued)

Table 6.1 (Continued)

Country	Programme profile	Year of adoption
	• Improving sustainable agricultural practices under increasing heat stress in the *River Nile* State	
	• Environmental conservation and biodiversity restoration in northern *Kordofan* State as a coping mechanism for rangeland protection under conditions of increasing climate variability	
	• Adapting to Strategies to adapt to drought-induced **water** shortages in highly vulnerable areas in Central Equatorial State	
D.R. Congo	• Energy-related projects	2006
	• The strengthening of agricultural production capacities: Multiplication of improved seeds of corn, rice and cassava	
Uganda	• Biodiversity conservation and restoration of Mangroves Marine Park	2007
	• Community Tree Growing Project	
	• Land Degradation Project	
	• Strengthening Meteorological Services	
	• Community Water and Sanitation Project	
	• Water for Production Project	
	• Vectors, Pests and Disease Control Project	
	• Indigenous Knowledge and Natural Resources Management Project	
	• Climate Change and Development Planning Project	
	(Note: contrary to the NAPAs of the other Nile countries, Uganda has defined generic but not specific priority areas)	
Kenya (not NAPA, but National Climate Change Action Plan)	• Long-Term National Low Carbon Climate Resilient Development Pathway	2012
	• Enabling Policy and Regulatory Framework	
	• Adaptation Analysis and Prioritisation	
	• Mitigation and nationally appropriate mitigations action	
	• National Performance and Benefit Measurement	
	• Knowledge Management and Capacity Development	
	• Finance	
	(Note: these are the nine sub-components of the Action Plan for the period 2013–2017)	
Egypt (not a NAPA, but a National Communication		1999

Table 6.2 Vulnerable sectors prioritised in the NAPA and national communication

Vulnerable Sector	Nile Basin Country								
	Congo	Egypt	Ethiopia	Kenya	Tanzania	Uganda	Rwanda	Burundi	Sudan
Agriculture	X	X	X	X	X	X	X	X	X
Water Resources	X	X	X	X	X	X	X	X	X
Human Health	X	X	X	X	X	X	X	X	X
Forestry	X			X	X	X			
Livestock				X	X	X			X
Energy			X		X	X			
Physical Infrastructure	X	X		X				X	X
Tourism		X		X	X				
Coastal Zone Management	X	X							
Aquaculture and Fisheries		X		X					

(Source: UNEP 2010)

initiatives has jumped the queue of political priority (UNEP 2010). The nexus between climate and transboundary water agenda has not always been highlighted but it is becoming increasingly important. Examples on how climate change issues are becoming 'regionalised' in the Nile/Horn of Africa region can already found (see below), but its direct link with Nile waters is just taking off.

Regional institutions

Although climate change action plans are expected to be developed and implemented at the national level, there is as well as increasing concern with the topic at the regional levels, with several of the institutions pushing the envelope for regional climate approaches. For example, the EAC and LVBC (respectively, the REC and RBO for the East Africa region) have been internalising climate issues in their activities, and even have adopted a common position on climate change negotiations (Jarso 2011). In 2011, the EAC Secretariat adopted a Climate Change Policy (EACCCP), and launched a joint five-year programme on Climate Change Adaptation and Mitigation and a Climate Change Master Plan (EACCCMP) for the region to supplement on-going efforts of the member states (five Nile riparian states) to cope with climate change, in which 'water' is also defined as a priority issue. LBVC, as a specialised institution of EAC, is operating under the same approach; and it has recently initiated an Integrated Climate Change Adaptation and Mitigation Programme (LVBC 2013). In 2003, the heads of state of IGAD established a specialised institution called Climate Prediction and Applications Centre (ICPAC), the mandate of which is to provide timely climate early warning information and support specific sector applications

Table 6.3 Priority adaptation projects in the Nile basin countries

Priority Adaptation Project Focus	Country								
	Congo	Egypt	Ethiopia	Kenya	Tanzania	Uganda	Rwanda	Burundi	Sudan
Water resource Management		X	X	X	X	X	X	X	X
Promotion of Drought-tolerant crops		X	X	X	X	X	X	X	X
Sustainable Agriculture and Land Management	X	X		X	X	X	X	X	X
Environmental Conservation and Biodiversity/Land Restoration	X		X		X	X		X	X
Early Warning Systems			X			X	X	X	X
Diversification of Energy Sources				X	X		X	X	X
Malaria Control	X		X	X	X		X	X	X
Integrated Coastal Zone/Flood plain Management	X	X		X				X	X
Strengthening Community Awareness	X	X	X	X	X		X	X	X
Livelihood Diversification	X		X	X			X		X
Water and Sanitation						X	X		
Indigenous Knowledge						X			
Disaster Risk Reduction/Risk Transfer			X	X					

(Source: UNEP 2010)

(disaster risk reduction products) to enable the region to cope with various risks associated with extreme climate variability and change (IGAD 2003), which includes all Nile riparian countries except Egypt.

In the same vein, the Nile basin initiative and its subsidiary institutions (ENTRO and NELSAP) have also been increasingly incorporating climate change in the programmes and projects. Because the NBI includes all the Nile riparians, the focus of this chapter will be mainly on the NBI regional/trans-boundary responses to the GCC discourse. Whether this is a case of adaptation, resistance, or subversion of the climate change discourse at the transboundary level in the Nile basin will be discussed in the next sections and the conclusion.

Analysis of the responses to GCC in the Nile basin

Contrary to the other chapters in this book, this chapter dedicated to the Nile does not aim to look at the national responses to climate change by individual riparian states. First, because it would be a time-consuming exercise, taking into account that 11 riparian states would need to be analysed. Second, that exercise would not necessarily bring any original contribution to the critical and original analysis this book is trying to put forward – i.e. how is climate change being 'tackled' at transboundary level in the basin. With this is mind, this chapter will focus primarily on response to climate change by the most inclusive (inclusive in the sense of including all riparians, and not only some of them) transboundary water institution in the Nile basin – the NBI. The analysis will allow under-standing of how the emerging GCC discourse has been affecting the TWM in the Nile basin, and how it has influenced the NBI institutions and how they have been incorporating climate change in its activities, programmes, projects, and discourse. But even before we move to that analysis, it is important to review what the main issues related to climate, which influence the discourse and action in the transboundary context in the Nile basin.

Climate variability vs. climate change in the Nile basin

Natural climate variability

Climate variability is a different concept from climate change. Climate variabil-ity in the Nile basin it is a fact, not just a prediction. In the Nile basin, climate is naturally variable. All the sub-basins of the Nile River system are characterised by climate variability, both intra-seasonal (annual variation) and inter-annual fluctuations (seasonal variation) in climatic parameters. Rainfall parameters show a high level of inter-annual and even inter-decadal variability (both in the Eastern and Equatorial sub-basins). The Nile flows are intimately determined by large-scale rainfall variability in the headwaters of the river, and runoff response amplifies variability (Conway 2005). Variable rainfall patterns are directly linked to flow extremes and climatic extreme events such as droughts and floods in the Horn of Africa region – the droughts of the 1970s and 1980s, and the floods in the 1990s, are recent examples of it (Goulden, Conway, & Persechino 2009). The Nile region is particularly affected by the climate variability because of a conjugation of natural and socio-economic factors that increase dramatically the vulnerability of the countries and its populations. Among these factors are: 1) high fragility of the natural systems two-fifths of the basin consists of arid and semi-arid dry lands); 2) dominance of poor and rural people, whose livelihoods (agriculture, pastoralism, fisheries) are highly sensitive to climate variability; 3) prominent agricultural sector downstream, very reliant on irrigation with Nile water; 4) rapid growing population; 5) high exposure to natural disasters; and 6) high sensitivity to changes in annual precipitation (NBI 2012b).

Climate change

Beyond the 'natural' climate variability, it is very likely that the Nile region will become more vulnerable to global climate changes. The scientific studies and models available identify the main impacts to be faced by the Nile countries, namely: 1) increase/reduction in river flow; 2) changes in the inter-annual variability of the flow; 3) increase in frequency of flood and droughts; 4) sea level rise in the Nile Delta ; 5) increase in surface temperature; 6) increase in evaporation; 7) changes in patterns of rainfall, runoff and sediment yield; 8) change in ecosystems and vegetation cover as a result of warming; 9) change in crop yields; 10) increase in reservoir evaporation rates; and 11) increase in evaporation rates from swamps.

For the purposes of this article, we are particularly interested on climate change impacts on the Nile water resources, in particularly in what concerns increase or decrease of runoff and water availability. Table 6.4 provides a summary of the main studies that have been conducted in the past 20 years on the impacts on Nile water resources. The summary reveals how the studies are indeed very divergent, showing that projections range from increase in the runoff (slight to significant) to reduction (significant to slight), and how some predict no major changes to the current situation.

Climate and water scenarios for the Nile basin

Climate and water

What is interesting to observe is that the several climate models for the Nile basin region clearly show that surface temperatures will rise, but the predictions in terms of river flows/water availability is not so clear cut. The several scientists that have been working on the climate models for the Nile basin, and looking in particular at the climate change implications in terms of rainfall, water availability, and evapotranspiration, consider that climate change global models are not precise enough to take conclusions for the Nile region and that only by downscaling climate models it will be possible to have good information to inform decisions and policies at the basin and sub-basin levels (source). Due to the size and complexity of the Nile hydrological and climatic system, there are some initial assumptions that need to take into account: 1) different countries will be affected differently; 2) different sub-basins will be affected differently; 3) arid and semi-arid countries are amongst the most vulnerable to the impacts of climate change (Jeuland 2010). Table 6.4 summarises the main findings of climate models applied to the Nile basin.

Table 6.4 Summary of studies on climate change impacts on surface water resources in the Nile basin region

↑↓→	Projected changes in water resources	Authors
↓ or ↑	Divergence between climate models results for the Nile Basin: two produce increases and two produce decreases in flows	Strzepek and Yates 1996
↓ or ↑	Range due to differences between GCM scenarios) of-9% to +12% changes in the mean annual Nile flows for 2025	Conway and Hulme 1996
↓ or ↑	By 2050, the coined effects of three driving forces (climate change, land-use change, water resources management) on future water availability in Egypt range from a large water surplus to a large water deficit	Conway et al. 1996
↑↑	Five out of six climate models produced an increase in Nile flows at Aswan	Yates and Strzepek 1998
↓ or ↑	Lake Tana: if temperature is increased by 2° and: 1) No change in rainfall > decrease in annual flow by 11.3% 2) Decrease in rainfall by 10% to 20% > decrease in runoff by 29.3% to 44.6% 3) Increase in rainfall by 10% to 20% > increase in runoff by 6.6% to 32.5%	Tarekegn 2000
↓	By 2025, propensity for lower Nile flows (in eight out of eight scenarios)	Strzepek et al. 2001
↑	Runoff in Eastern and Northern Africa is projected to possibly increase by 2050	Arnell 2003
↓↓	Significant reduction in runoff in Nile by 2050	Manabe et al. 2004
↑↑	Increase in runoff of 20 to 40% by 2050 in Eastern Equatorial Africa	Milly et al. 2005
↑	Runoff in Eastern Africa is projected to possibly increase by 2050	Strzepek and McClusley 2006
↑ or →	Future Nile discharge (up to 2100) will decrease slightly (−2%) or will remain relatively stable to the current situation	Aerts et al. 2006

(Source: adapted from Goulden, Conway, & Persechino 2009: 811)

Scenarios

The extant studies on climate change and water in the Nile basin show how different and divergent the conclusions are, as well as the different sub-basin specificities one has to take into consideration. But basically what can be observed is that there is no consensus about the increase/decrease of rainfall and increase/decrease of water availability – it might not change (*status quo*), it might increase and it might decrease. In any of the scenarios, what is clear is that whatever will happen, it will not be only (and not even mainly) influenced by natural and climatic factors, but it will be mainly influenced by social, economic, and political factors. Therefore, any analysis to be carried on the climate change impacts must be combined with non–climatic factors that might be more influential for flows availability than the natural factors themselves. Some of the important factors to take into account are: population growth and density, land use management, agricultural practices, consumption trends, energy policies, economic development options, etc. (Jeuland 2010). In brief, we are in presence of a catch-22 situation that we need to carefully take into account when analysing how climate change is being factored in national or transboundary arrangements, both in discursive and policy terms. The catch-22 paradox can be expressed in very simple terms: On the one hand, climate change can influence economic/agriculture/energy policy options of countries and regions (e.g. it can lead national governments or regional institutions to move towards an agriculture based on drought-resistant crops or to adopt a certain infrastructure planning in preparation for extreme events such as floods). But, on the other hand, the economic/agriculture/energy policy options that countries and regions will be taking in the short and medium term will also influence the impacts of climate change (e.g. the construction of a large-scale dam or increase on irrigated agriculture can have both positive and negative impacts on river runoff and sea level rise). In the concluding section, this chapter look at the particular nexus between energy, agriculture, and climate agendas, an area of specific prominence in the case of the Nile.

Addressing climate uncertainty and change at national and regional levels

National level

NAPAs are being mainstreamed into existing programmes and national development programmes. Some countries have prepared NAMA programmes (e.g. Ethiopia Climate Resilient Green Economy Initiative). Some countries established special climate change units (e.g. Kenya National climate change Activities Coordination Committee). Climate change adaptation is mainly about socioeconomic development, and, as such, interventions need not be viewed as standalone set of measures, but rather as integrated parts of the existing development programmes and policy initiatives. In any case, it is widely recognised by many authors that the scope of the current climate change adaptation programmes and activities in the Nile countries is not sufficient to deal with the scale of the threat. There is a generalised consensus that country efforts in adaptation actions should and must be complemented by mainstreaming climate change at regional level with a transboundary river basin perspective. The question is to what extent is this already occurring at the moment, and with what degree of success.

Regional/transboundary level

Taking into account such level of uncertainty concerning climate and water in the Nile basin, the way of addressing climate change becomes a complicated puzzle not only at a national but as well at a regional level. If national economic/

agriculture/energy policy options at national level will have implications, also the levels of transboundary conflict and cooperation will. Goulden, Conway, and Persechino (2009) based on past case-studies of climate events, have tried to address two very important questions: 1) What were the responses to these events and what was the role of the regional institutions to these responses?; and 2) How has cooperation/conflict influenced adaptation? The same authors also look at the future and ask questions the following questions: 1) Is cooperation between riparians necessary for adaptation to climate variability and change?; and 2) Will the threat of climate change in the future prompt cooperation between riparians? The conclusions of their study highlight that transboundary cooperation will be essential for the countries to develop efficient strategies to adapt to climate variability and change – and inter-governmental institutions such as the NBI will have an important role to play. As the next section shows, the NBI and its institutions have indeed already started working on the adoption and internalisation of a climate agenda.

Nile basin initiative: when climate becomes transboundary

The NBI approach to climate change

If we take the NBI as our unit of analysis to understand how a regional/transboundary institution is addressing climate uncertainty and change, it is possible to observe that climate is definitely being taken up as a priority issue, at least in discursive politics and policy papers. The vision of the NBI clearly states that the water sustainability in the Nile basin has to take climate change into account; and the equitable and reasonable utilisation of the Nile water resources also has to take climate change into account. Ultimately, a key feature of the NBI mission is 'the capacity and necessary mechanisms for the countries to take a regional approach to minimise climate change threats to socio-economic growth and development' (NBI 2012). Several of the NBI, ENTRO, and NELSAP programmes, activities, and tools put this in evidence. As the box below shows, since its inception, and not only in the last years, the NBI had been working on several fronts related directly or indirectly with climate variability and change in the basin.

Current and past NBI activities contributing to climate resilience

1   **Bridging the knowledge gap:** promoting studies at the basin and sub-basin levels that include climate change factors and appropriate coping mechanisms – e.g. Cooperation Regionals Assessments (CRAs) in ENTRO, MSICA in NELSAP, basin-wide climate change assessments by the Shared Vision program Water Resources Planning and Management (WRPM);

2   **Strengthening monitoring and planning tools:** hydrometero-
logical monitoring networks (NELSAP), Flood Forecast and Early
Warning System (ENTRO), Nile basin – Decision Support System
(DSS) basin-wide planning tool (WRPM);

3   **Supporting science-policy dialogue:** Nile Basin Development
Forums (last one specific on CC), State of the Basin Report, trainings;

4   **Facilitating expansion of the region's water and power infrastructure:** through the investment projects of ENTRO and NELSAP
that now also include climate-proofing approaches;

5   **Promoting watershed management:** NTEAP (micro-projects),
and at ENTRO and NELSAP;

6   **Promoting idea of transboundary level adaptation measures:**
based in the idea that climate change adaptation will be most effective when undertaken in coordination with other riparians; a set of
'no-regret' measures is suggested by the NBI.

No-regrets approach

The NBI approach to climate change is based on a 'no-regret' approach based on
a series of measures that should be incorporated in the projects being developed
at regional (but also national) levels. These measures include:

• Increased water storage capacity
• Interconnection of electricity grids
• Land-use planning
• Expanding forests and reverse deforestation
• Increase capacity for over-year food storage
• Infrastructure for intra-basin agricultural trade
• Increase productivity and water-use efficiency in irrigated agriculture
• Mitigation impact of drought
• Mainstream climate change adaptations and mitigation plans
• Increase research
• Build capacity at national and national levels
• Developing mechanisms for soliciting climate change adaptation funds
• Operating joint hydrometerological programmes

Among the several 'no-regret' measures, the NBI but considers that the priority measure is to expand water storage infrastructure (large and small) in the Nile
region (NBI 2012b). This priority had been established based on a series of assumptions. First of all is that water storage (mainly upstream) is the basis for assuring
water productivity in the face of threats of climate change. Second is that water storage is necessary for mitigating economic effects of the already-existing hydroclimatic
variability. The NBI assumes as well that, for the moment, the Nile region countries

have insufficient water storage capacity, and that any planned water storage needs to be part of any balanced water-investment programme, such as those being promoted by the NBI. In what concerns large dams projects within the basin, the NBI considers that they are best located in regions with low evaporation rates, and will require levels of cooperation and clear mechanisms for sharing costs and benefits. Despite the fact that this idea seems to be consensual among the technical communities, it is known that is politically charged. Egypt is still very much resistant to the idea of moving storage upstream, due to the current limited trust and confidence between neighbouring riparians. Another priority of the NBI is promoting specific knowledge about climate change impacts in the region, by downscaling from global models and predictions. The goal is to promote understanding on the future climate in the basin and act against uncertainty. More hydrometerological monitoring (stations and comprehensive data network) is needed in order to have more accurate and robust climate/water information systems.

All of the arguments presented in the above paragraphs are summarised in *The State of the River Nile Basin* (NBI 2012), a document that took several years of conceptualisation and negotiations between the different Nile riparian countries, that into account that the NBI aimed at producing a document as clear and consensual as possible. The atlas also reflects the NBI Statregic Plan of 2012.

Operationalionalisation

The NBI is already mainstreaming climate change in the NBI activities in several different ways. On the one hand, in the last decade the NBI institutions had been already incorporating climate change in the project preparation and implementation phases. Almost all the pre-feasibility and feasibility studies for the SAPs (NELSAP and ENSAP) investment projects have included climate change components. On the other hand, in 2012 the NBI has adopted an Environmental and Social Management Framework, which is part of the Nile Basin Climate Resilience Growth Program (NBCRG Programme), which informs all current and future NBI activities, programmes, and planned infrastructure. The development objective of the NBCRG is 'to improve climate change resilient water resource management and development in the Nile Basin' (NBI 2012b). Two of the expected programme's outputs are: 1) to develop a portfolio of climate resilient catalytic transformative investment projects, and 2) to develop a knowledge base and analytical framework for climate-resilient planning. The ultimate goal of the NBCRG is to promote coordinated and optimised climate proofed planning of water resources.

Regional vs. national projects

One of the main challenges of the regional approaches to water management and climate is related to the current legal and institutional pitfalls of the NBI. The NBI is still a transitional arrangement and not yet a permanent organisation with full legal status. And at the same time, Nile countries have been developing hydraulic infrastructures unilaterally or bilaterally, without necessarily having taken into account all the environmental (including climate) standards as defined at the regional level. Unilateral developments can potential limit the ability of cooperation to contribute to climate change adaptation, in particularly because they might well fail in addressing the climate complexities in a holistic manner, as the national focus might prevail at the expenses of a regional understanding of the climate risks and possible adaptation and mitigation strategies. But due to known hydropolitical intricacies of the Nile basin, one might have to assume that there is still a long way to go until the NBI (or the future Nile Basin Commission) can translate ideal regional frameworks into national practice. Nevertheless, the NBI institutions are already paving the way for a regional thinking whereas climate-proofing conceptualisations are included, as the ENSAP examples (discussed next) shows.

Example of the Eastern Nile/ENTRO

The Eastern Nile Technical regional Office (ENTRO), the NBI subsidiary institution for the water management and development of the Eastern Nile Basins, has adopted climate change as an important element of its operational directives. In 2008, it conducted a *Study on Climate Smart/Proof for the ENSAP Projects* – the study identified the climate components to be included in several of the projects, namely the Watershed, Irrigation, and Hydropower investment projects. In 2009, the *ENTRO Approach Paper* delineated a comprehensive strategy on how to respond to the challenges and opportunities associated with climate change, and has defined five pillars: prediction, adaptation, mitigation, potential opportunities, and education. The *Study* (ENTRO 2008) and the *Approach Paper* (ENTRO 2009) have set the stage for a coordinate incorporation of climate change components within each of the ENSAP projects, which are now based on 'climate-smart' concepts, studies, and activities.

In 2010, the ENSAP *Climate Change Proofing Action Plan* was concluded (see Development Briefs and Concept Notes), which establishes the guidelines for climate proofing projects (ENTRO 2011). The Multipurpose Baro-Akobo-Sobat Project (BAS), to be financed by the African Development Bank, is the ENSAP pilot and show-case investment project in what concerns climate change adaptation and mitigation (ENTRO 2013). This Baro-Akobo-Sobat sub-basin, shared between Ethiopia and South Sudan, is of particularly importance taking into account that it crosses the Sudd region, the biggest African wetland.

Although the BAS Project will be the first multipurpose project to fully adopted a climate-proofing conceptualisation, ENTRO already has the experience with the Watershed Management Project (involving Ethiopia and Sudan), that has included several components of mitigation of and adaptation to climate change, in particular by investing in land-use management (ENTRO 2007). The project aims at preservation of land cover and reduction of loss of vegetation cover. The baseline is that change in land utilisation and management is one of the processes driving global climate change. Therefore, any efforts towards reducing the rates of change in land cover should also help in reducing effective emissions of carbon. The WMP Fast-track project indicated that is possible to quantify its contribution in addressing climate change mitigation via enhancing preservation of land cover and reducing loss of vegetation cover. This project had been financed partially through global climate change financial mechanisms.

NBI and the nexus water-energy-climate

The NBI and the two subsidiary institutions (ENTRO and NELSAP) had been strong in advocating and promoting energy agendas considered to be climate proofing, namely in its ambitious to develop hydropower dams in the Nile tributaries. The Joint Multipurpose Project (JMP) and Eastern Nile Power Trade (ENTP) projects in the Eastern Nile, and the Nile Equatorial Power Pool in the Nile Equatorial basins are examples of it. The baseline assumption is that power market integration can be a climate change mitigator, as integration of the national and regional markets of energy can facilitate engagement with CDM. Basically, by producing energy in the upstream catchments of the Nile basin, it can contribute to decreasing the power generation in the downstream riparians – and increase carbon offsets and reduce reservoir emissions. At the same, the same projects may provide adaptation benefits to the whole region.

Energy demands are immense and increasing in the entire Nile basin region, and institutions like the NBI have tried to answer the question on how to meet these demands. The NBI, ENTRO, and NELSAP have developed studies where a combination of options, including hydropower (power development plans for each country and for the region), could be financed through CDM funds (the CERs) in order to promote the green renewable option of importing electricity instead of going for grey polluting options (example: Egypt/Sudan imports electricity from Ethiopia, reduces substantially its emissions by reducing the national natural gas/coal exploitation by importing electricity from Ethiopia). One of the main challenges that the NBI faces is that CDM engagement is with single countries and not a group of countries – but securing finance is possible if there is regional collaboration, and the NBI has been working on this resource mobilisation opportunities. The ideal scenario for example include Egypt/Sudan importing electricity from Ethiopia and applying for CERs, which could be translated not only in enormous energy and environmental gains, but as well in financial gains for all the countries.

The NBI has been working in the nexus Water-Energy-Climate, namely by promoting studies and developing the following tools:

- Potential of integration between national energy markets
- Analysis on how integration can offer an opportunity for a different screening criterion to be used in the selection of new energy sources in the region
- Analysis on how integration can be used within the CDM process to secure CERs

- Develop tools to calculate carbon emissions due to power development trajectories with and without the regional power trade options (and calculate the carbon emissions savings of the regional power market option)
- Roadmap carbon trade and financing engagement

Adapting, resisting and subverting the climate change discourse in the Nile basin

Looking at the TWM institutional responses to climate change debate in the Nile basin, it is possible to observe that there is an escalating trend for the inclusion of plans, policies, and projects addressing climate change issues. But there is a need to analyse and distinguish what are 'real' measures towards adaptation to climate change and what is merely a cosmetic operation for political or financial reasons. In the Nile basin, it is possible to find elements of all three 'strategies' identified in Chapter 1 of this book – adapt, resist, and subvert. The three coexist in order to serve the political, financial and/or socio-economic interests of the several actors, both at national and regional level. When looking in particular at the responses by the transboundary institutions (the NBI centres, in particular), the most dominant strategy is most likely the 'subvert', as explained next.

Adapting

In theory and on paper, all the Nile basin countries are already adopting or adapting national policies in order to strengthen the institutions to meet present and future climate change and variability challenges. To move from policy to practice is, nevertheless, a process that might take more time. The same can be said about looking at climate change as a cross-border and transboundary issue, and not merely as a national problem to be address by national governments. Nevertheless, the analysis shows that Nile countries through the TWM institutions, namely the NBI, are acknowledging that climate change and water-related climate issues as problems that need to be tackled jointly. Although most of it is a very discursive process, there are already clear examples – at the operational, system, and structure level – that show some adapting responses at the transboundary level, being put forward by the multilateral institutions.

Organisational

The climate change discourse is becoming a standard element of the operations of organisations directly or indirectly working with the management of water resources. Institutions at the national and regional level now employ staff that has expertise and skills in climate change, forecasting, early warning systems, adaptation and mitigation measures, etc. And the same organisations also regularly train their staff on climate change-related technical issues, raising awareness and increasing the capacity of the countries and institutions to deal with climate variability, change, extreme events, and their impacts. At the national level, most of the countries now have units at the Ministry of Water Resources (and other related ministries such as Energy, Environment and Agriculture) dealing specifically with climate issues. At the regional level, institutions such as NBI, ENTRO, and NELSAP, but as well the LVBC and IGAD, have specialised teams working with these issues, be it in specific climate-related projects or linking it with the existing soft and investment projects being carried out by those institutions. A visit to the institutional websites of any of these organisations will show how climate features prominently in their project portfolios. These multilateral institutions have also organised an immense number of trainings and workshops, namely on climate change funding mechanisms, at national and regional levels. In brief, one can assume that climate is now part of the project cycle – from its design to preparation stages, and even to the implementation stage (when and if the projects will ever reach that stage).

System

Almost of the Nile countries have in place early warning systems that aim at increasing the awareness and coping strategies of the countries to deal with climatic extreme situations, that include tools such rainfall observation, drought forecasting, and modeling applications that can be used to prevent major humanitarian crises when information gathering is well linked to policy-making decisions (Maxwell & Watkins 2003; Verdin et al. 2005). For example, Horn of Africa countries have sophisticated regional early warning systems in place that in the last years have helped the countries in the preparation, response, and recovery in cases of drought that frequently hit the Horn of Africa region. However, the degree of efficiency varies a lot – this was particularly visible in the severe drought events of 2010 and 2011, when Ethiopia and Kenya have prevented major humanitarian crisis in their countries but the same have not happened in the neighbouring Somalia, which shows that to have the climate information available is not the only factor that matters (Hillbruner & Moloney 2012; Ververs 2012). At the bilateral level, Ethiopia and Sudan have in place national early-warning systems that foresee the exchange of information about the changes in rainfall levels and acute cases of floods. Nevertheless, the level of prevention and response has not been as efficient as expected. The extreme impacts of floods in May 2013 in Sudan, with dozens of people killed and displaced, once again show that having early warning systems in place might not be enough to increase the resilience of countries and populations to extreme events (Walker 2013).

At the transboundary level, there was been as well good attempts to bring to life early warning systems with a transboundary nature – based on the assumption that climate events in one riparian country also affect the other, and that joint action can bring benefits to all the countries. ENTRO's Flood Preparedness and Early Warning System (FPEWS) is a good example of how the NBI is moving ahead with an 'adapting' transboundary institutional response, but how this is only the beginning of the learning curve, as discussed next. FPEWS aims at promoting flood risk management and non-structural approaches to managing the impact of flood in the Eastern Nile region; and to enhance regional collaboration and improve national capacity in the mitigation, forecasting, warning, emergency preparedness, and response to floods in the Eastern Nile basin (ENTRO 2014). This programme is an example of a transboundary institutional response, but also shows that there is still a long way to go to promote resilience to climate change at the transboundary level. The programme so far has implemented pilot projects that put in evidence the positive benefits that working jointly can bring: increased information availability exchange of expertise between countries, preparations of initial steps of a flood and forecasting systems and modeling, awareness creation, etc. (ENTRO 2014). In-depth interviews with ENTRO staff and officials from Ethiopia and Sudan (the two countries involved in this project) brought to the forefront important messages and lessons to be considered by all NBI centres and countries: there is still a lot of work to do if climate change is to be addressed in an efficient manner at transboundary level. Some of these lessons are: more needs to be done in terms of institutional strengthening of national and regional institutions in a continuous basis, systematic information sharing between countries needs to be encouraged and incentivised, and technical and institutional coordination between neighbouring countries needs to be consolidated. For this to become a reality, Nile riparians might need to move a step forward and commit more financial and human resources to upscale programmes such as the FPEWS and above all reinforce political commitment, i.e. signs of strong political to move transboundary adaptation responses from paper/policy to real action on the ground.

Structure

Development of hydraulic infrastructure as a measure to adapt to climate change impacts has been on the agenda of both the Nile countries and the transboundary institutions. Looking at the national adaptation plans, as well at the NBI investment projects, the development of infrastructure such as storage reservoirs for both flood control and mitigation measures in periods of drought it is envisioned. Examples of it can be found for example in the Ethiopia's Climate Resilient Green Economy Initiative, a strategy that is part of Ethiopia's ambitious five-year Growth and Transformation Plan, and that includes the development climate-proofing infrastructure projects in the Agriculture, Forestry, Power, Transport and Industry sectors – including a strong emphasis in the development of hydropower as a clean and renewable energy source (Ethiopia 2011). At the basin level, the Nile Cooperation for Results (NCORE) Project, which is basically the current financial stream supporting activities under the Nile Basin Climate Resilient Growth Program, put an emphasis in supporting climate-proofing investment projects in the two Nile sub-basins, although the feasibility of doing so in the short or medium term might be remote, taking into account the current political situation in the basin.

However, less has advanced in terms of structure level as an adapting strategy, than in terms of operational and system levels as analysed above. Interviews conducted with decision-makers from most of the Nile riparians and NBI institutions indicate three of the fundamental reasons why climate-friendly infrastructure planning will take long to be taken over at the transboundary level. First of all, there is not yet enough regional political buy-in in the Nile basin to move forward with the implementation of joint and integrated infrastructures, even if all the benefits have been identified and understood by the riparian states. Second, because those transboundary infrastructures require massive financial resources that are not immediately available, and the financing processes through traditional donors parties are being overcome by easier and faster funding (usually for unilateral projects) from non-traditional financiers (Cascão 2009; Nicol & Cascão 2011). Third, the issue of building infrastructure (namely storage) upstream, where the potential for climate-related benefits could be achieved, continues to be extremely controversial and politically sensitive in the Nile basin region. In brief, the Nile riparian states are still planning infrastructure development too much from a national perspective, and the NBI is still lacking institutional and legal clout that allows the organisation to play a more active role in potentially implemented investment/infrastructure projects that the NBI institutions have been identifying in the past decade.

Resisting

As mentioned earlier in the chapter, climate variability is a constant in the Nile basin. The history of the Nile is made of uneven levels of rainfall and climate extreme events and as such it is a fact difficult to be ignored by any of the countries. Major headlines in the global media covering the devastating impacts of droughts and floods in the Horn and East Africa regions for the past four decades make it even more difficult to deny that climate is a relevant political factor in the region. The existing models, predictions, and forecasts for climate change and future rainfall variability in the basin, although pointing in several different directions (see previous section in this chapter), are difficult to be ignored by the countries and the transboundary institutions. In brief, climate variability and change are such obvious elements in the Nile basin that it is almost impossible to ignore or resist it. Examples of adapting/subverting climate change discourse (be this discourse at global or at basin level) are the most common strategies in the Nile basin (as analysed in this section), but in some specific cases resisting strategies can also be found – not resisting in terms of avoidance of issue, but resisting by dealing with climate issues in a politically-motivated selective manner. Resistance to climate change discourse in the Nile basin can be spotted in two different ways: 1) in the legal responses to climate change, and 2) in 'cherry-picking' type of strategies.

Legal agreements

Not surprisingly, the existing legal agreements in the Nile basin do not address directly the issues of climate change, although they include allusions to possible collective responses to climate/hydrological extreme events, but that is done in a very selective manner. This chapter looks at how the two major and controversial agreements in the Nile basin – the 1959 bilateral agreement on the Nile waters between Egypt and Sudan, and the 2010 Cooperative Framework Agreement (so far) signed by the upstream riparian states.

The 1959 Agreement on the Full Utilisation of the Nile Waters does not mention climate change explicitly, which is normal because in the 1950s this was not yet a concept in vogue, but nevertheless it does recognise the existence of climate/hydrological variability in the basin and even included provisions on how to deal with 'low years', i.e. years that the water flows might be lower than the average in result of drops in the rainfall levels upstream. Article 4 of the Agreement says:

Article 4. Technical co-operation between the two republics

'As it is probable that a series of low years may occur, and a succession of low levels in the Sudd el Aali Reservoir (Aswan High Dam reservoir) may result to such an extent as not to permit in any one year the drawing of the full requirements of the two Republics, the Technical Commission is charged with the task of devising a fair arrangement for the two Republics to follow. And the recommendations of the Commission shall be presented to the two Governments for approval.'

What is interesting to observe in this case is that the 1959 Agreement acknowledges the existing climate/hydrological variability and even advances with an institutional response to it, which is something unusual in international water agreements. But two fallacies can be found in this approach. First, the 1959 Agreement only refers to the 'low years', but fails to mention anything about the years when the flows are above the average determined in the agreement, although it is know that the annual average of 84bcm is exactly that – an average between 'low' and 'high' years. This puts in evidence that the 1959 Agreement only reflects the concerns of the two downstream countries, that are mainly concerned with possible impacts of decreased water availability and how it could affect the absolute (measured in billion cubic metres of water and not percentages) water allocations provisioned by the Agreement. Second, the 1959 Agreement deliberately avoids mentioning the 'high years' when there will be surplus of water (and there had been many years of high flows, according to several scientific studies), so that these waters remain unaccounted and can be 'silently' used by the two downstream riparians (see for example Waterbury 1979, 2002). This comes in line with the fact that the other Nile riparians were not part of the Agreement, and are not considered in any possible response and/or prevention measures that could be taken collectively (be it in the case of high or low flows), although the other riparian countries are the sources of all the water reaching Lake Nasser.

The 2010 Cooperative Agreement, being a more recent legal document, already incorporates several of the 'new' language adopted by international framework conventions and water agreements, and it explicitly refers to climate-related impacts (Nile Cooperative Framework Agreement 2010). Although the CFA does not mention climate change as such, it refers to it in the article about 'emergency situations', in its Article 12.

Article 12. Emergency situations

1   *For the purposes of this provision, 'emergency' means a situation that causes, or poses an imminent threat of causing, serious harm to Nile Basin States or other States and that results suddenly from natural causes, such as floods, landslides or earthquakes, or from human conduct, such as industrial accidents.*
2   *A Nile Basin State shall, without delay and by the most expeditious means available, notify other potentially affected States and competent international organizations of any emergency originating in its territory.*
3   *A Nile Basin State within whose territory an emergency originates shall, in cooperation with potentially affected States and, where appropriate, competent international organizations, immediately take all practicable measures necessitated by the circumstances to prevent, mitigate and eliminate harmful effects of the emergency.*
4   *When necessary, Nile Basin States shall jointly develop contingency plans for responding to emergencies, in cooperation, where appropriate, with other potentially affected States and competent international organizations.*

Interesting to observe in the language adopted in this provision is that several of the steps of the 'CC adaptation chain' are taken on board – prevention, preparedness, response, and recovery. The CFA reflects the current times by adopting the sophisticated GCC discourse. And it is also pioneering in the sense that indicates that the response to climate extreme events must be collective, and not only at national level. This can be considered as a clear TWM institutional response to climate change in the Nile basin, although it cannot be ignored that the CFA is not yet a binding agreement, as countries are still in the process of ratification and the two downstream riparians are not willing to sign it (mainly because of political reasons already mentioned at the beginning of this chapter).

But what is surprising in this Article 12 is the fact that droughts have not been included as an 'emergency situation', although they are shortly mentioned in another article (Article 11) of the CFA. Knowing that the Nile upstream riparians have been affected by several droughts in the last four decades, it is strange that droughts have not been included in this article. In particular because it is known that there is a direct link between drought events upstream and decrease in runoff and water availability in the Nile hydrological system. It appears as if this is the other side of the coin of the 'cherry-picking' strategy of the 1959 Agreement, as analysed next.

'Cherry-picking' strategy

The above discussion about the legal response to climate change shows that the two existing legal agreements (specific or framework) that regulate the Nile basin do not incorporate climate change as such, although they mention climate variability in a selective way. This can be considered as sign as to how the climate change discourse can be used (or it has been used) as a political card in the Nile hydropolitical game. Moreover, as climate change models for the Nile basin show so many contradictory trends – some indicate an increase in water availability, some others a decrease – the conclusions of the studies can also be used by the several actors according to their own interests.

Looking at the past, there are several examples on how climate has been blamed for major crisis in the countries and to minimise responsibility of decision-makers for the lack of prevention and response to the disasters in order to avoid humanitarian crisis, and major negative impacts to its populations. For example, during the 1970s and the 1980s, the responsibility for major food crisis and famines in the Horn of Africa and consequent death of millions of people was attributed to the droughts and it was convenient excuse to hide wrong and manipulative policies by national governments (de Waal 1997). The case of Ethiopia is a well-known one, when the governments of Haile Selassie and Mengistu, respectively in the 1970s and the 1980s, failed to recognise that the massive famine was not only an outcome of the droughts, but as well of ill agricultural and development policies, wrong food aid distribution, and political motivations related to punishing rebels and opposition groups (de Waal 1991; Gill 2010). Several also accuse the current Ethiopian government of using similar arguments – blaming the uneven rainfall patterns – to justify the 'green famine' affecting the southern regions of the country (Piguet & Bawtree 2003; Handino 2014). But it is not only in the cases of drought that climate change has been the perfect excuse for political mismanagement, also in the case of floods during the 1990s and already in this century we can see governments of countries like Sudan and Ethiopia using the climate card to minimise their political responsibility to prevent negative impacts on their most-vulnerable populations.

More recently, Ethiopia and Sudan as well as other upstream countries are using the climate discourse in a less-apologetic and rather a more-constructive way – namely suggesting that infrastructure upstream is urgently needed in order to prevent disasters like those in the past. By highlighting that rainfall patterns are uneven in the region and that climatic extreme events are becoming more frequent, they are putting high in the agenda the construction of new dams, reservoirs, and other water storage facilities as ways of mitigating and adapting to the climate variability and change, not only in their own countries but in the basin as a whole (see for example, Ethiopia 2011). The argument is not new but it is now more widely accepted among the global community exactly because of the climate argument, and it is getting a lot of support by international and regional players (see AfDB African Development Report 2013).

Interestingly enough, Egypt uses the card of climate change to justify sea level rise and soil salinity in the Delta areas, which are identified as major CC-related threats for Egypt (e.g El-Raey et al. 1999; Sherif & Singh 1999). But it is rare to see Egypt acknowledging that part of the problem in the coastal areas are a direct result of the agricultural/water policies of the last decades, which consists of over-utilisation of the Nile waters and the release of almost nil freshwater to the Mediterranean Ocean (Conway 1996). On the other hand, Egypt resists the use the climate change discourse when it puts forward the argument that upstream countries should use their 'immense' green water resources instead of insisting in utilising the Nile surface waters (Abu-Zeid 2008), but ignoring that rainfall and climate variability significantly affects the green water availability in the upstream catchments and the capacity of the upstream riparians to 'manage' that green water in a systematic way.

Subvert

Most of what can be observed in the Nile basin – national and regional level, in particular in the last couple of years, it is mainly subversion of climate change debate as a tool for political or financial purposes. As exposed in Chapter 1 of this book, by subvert it is meant the transformation of the initial meaning of the debate – the adaptation to climate change and mitigation of its impacts – in order to suit the interests of certain actors in specific situations. It can be said, for example, that the NBI for example has been using the climate change debate as a tool, as examples below show.

CC as smoke-screen for 'failed' transboundary cooperation

As analysed earlier in this chapter, in the past couple of years a lot of political weight has been given to the climate change within the context of the Nile cooperation process. One of the main reasons, as mentioned in the next point, is related to increasing the funding mechanisms for programmes and projects being executed by the NBI centres, but as well for other political reasons that can also be identified, namely because of the fact that the two tracks of the transboundary cooperation – NBI and CFA – have not been delivering according to the initial expectations (of countries and development partners) raised in the mid-1990s (Cascão 2012). A deadlock in the political track (the CFA) from 2010 onwards has affected the technical cooperation (NBI), in the sense that the new legal and institutional framework was not yet adopted, and a permanent river basin commission has not been established. As such, the NBI remains a transitional arrangement, investment projects are in standby situation, and a steady flow of finance from traditional development partners is still waiting for positive development in the political track. One of the ways out for this impasse has been the re-dressing the support to transboundary cooperation as support to climate change initiatives. The NBTF-supported

Nile Basin Climate Resilient Growth Program is a clear example of it. The financial support to the NBI is mainly directed to climate-related activities (World Bank 2012) – mainly soft projects, as financial support to infrastructure (in particular in the Eastern Nile region) keeps being hostage to major political decisions in the basin. Without a clear legal framework and permanent river commission in place, it is likely that financial support from traditional development partners to large-scale infrastructure in the Nile basin is not going to take place. The NCoRe project is financed from a grant from NBTF Grant in the amount of USD 13.8 million and a CIWA Grant in the amount of USD 1.5 million (NBI 2012a).

The 'move' towards support to climate change activities in the Nile also comes in line with major changes in the development agendas, in particular in European countries, which have been moving away from supporting politicised and securitised risky transboundary cooperation processes such as the current setting in the Nile and moving towards financial support to more ambiguous (and less-risky) agendas, such as climate change. Although this might be a mere re-labeling or cosmetic cooperation, the fact is that development partners are putting the climate change discourse high in their agendas, and transboundary organisations are deemed to adapt to these news discursive realities. The NBI appears once again the perfect place to test this new type of conceptualisations – as it was before with transboundary cooperation, benefit-sharing, and alike concepts – and it is already adapting the discourse, and even the projects preparation, to cope with this new sanctioned discourse.

Climate change as a source of financing

The adoption – discursive and in project preparation – of climate change as a keyword in the NBI operations comes not only as a reaction to the bilateral development partners' agenda, but because it is becoming more evident that global climate change mechanisms can provide financial resources that would not be otherwise available. The resource mobilisation strategies in the NBI centres are being re-strategised in a way that take into account the new streams of financial available by international organisations and international financial institutions (NELSAP 2011). This new trend is influencing not only the new projects being prepared by the NBI centres, but also the existing ones that are now being redesigned to become or look like climate-proofing projects, namely by including adaptation and mitigation strategies. NBI centres have been investing time and funding to train its technical staff on how to write proposals to apply for climate change funding mechanisms and to circumvent the very fact that climate change funding mechanisms are usually provided to government and not to regional/transboundary institutions. Although ultimately the projects might ultimately indeed contribute to the 'battle' against climate change impacts in the Nile basin, for the moment this appears more as subversion of the global climate discourse to serve specific financial purposes.

Climate change as a drive to return to large-scale infrastructure

Climate change provides a new opportunity for funding of transboundary cooperation, and it is not only for 'soft' projects dealing with technical and policy issues, but as well for funding of large-scale infrastructure. For example, there is now a new way of packing the investment projects such as large-scale dams that until recently were not considered politically correct after a longstanding debate about the negative impacts of such dam projects (World Commission on Dams 2000). The 'package' now includes arguments such as: multipurpose dams have a role to play in reducing the vulnerability to climate change, because they have capacity to increase the adaptation and mitigation of climate change impacts, by for example contributing to flood prevention and mitigation of droughts (see for example ENTRO 2011). Besides that, there is an extra-powerful argument being used: hydropower is a clean energy and, therefore, a climate change-friendly strategy (see for example Ethiopia 2011).

The Nile, which is still a relatively 'undamned' river compared to other large rivers in the world and where floods and droughts are a persistent problem, is the perfect setting to test the argument. The countries and the NBI have in the pipeline a dozen of such projects – hydropower/storage dams in the Blue Nile in Ethiopia and Sudan (including the Grand Ethiopian Renaissance Dam) and the multipurpose dams in the Equatorial Lakes region (the Rusumo Falls in Burundi/Tanzania, the Karuma Dam in Uganda). Therefore, the story goes that these large-scale projects will contribute not only to the socio-economic development and the water, food, and energy security of the countries and its neighbours, but will also contribute to the climate security in the region (NBI 2012b).

The above arguments might be sound, but scientists also call the attention for the fact that there is an urgent need to think about it in an integrated, systematic, and transboundary manner, which implies an harmonisation between all the dams and its reservoirs systems (and national authorities of the countries where the projects will be located) and the planning and management of the water storage, in order to avoid the opposite outcome, i.e. that climate change will 'constrain the technical performance of large reservoirs with knock-on effects for agriculture and electricity production'. The NBI was established exactly to perform this role of bringing the countries together to identify, prepare and facilitate the best and wiser *investment* in water development infrastructure *projects* in the basin. But the deep-rooted hydropolitical hurdles are preventing the institution from playing that role and pushing the countries to develop the infrastructures unilaterally instead of multilaterally, which might end up not being the best way of promoting climate-proofing best options.

References

Abu-Zeid, M., and Abdel-Dayem, S., 1992, 'Egypt's programmes and policy options for facing the low Nile flows', in Abu-Zeid, M. A., and Biswas, A. K. (eds.), *Climatic Fluctuations and Water Management*. Oxford: Butterworths and Heinemann: 48–58.

Abu-Zeid, K, 2008, 'Green water and effective legislation for transboundary water management', 89–95, Med 2008, 2007, in the Euro-Mediterranean Space, IE Med and CIDOB, SSN:1698–3068.

African Development Bank (AfDB) Group, 2013, 'African development report 2012: towards green growth in Africa', African Development Bank (AfDB) Group, Tunis.

Agrawala, S., and Van Aalst, M., 2008, 'Adapting development cooperation to adapt to climate change', *Climate Policy*, (2): 183–193.

Cascão, A. E., 2009, 'Changing power relations in the Nile River basin: unilateralism vs. cooperation', *Water Alternatives*, 2(2): 245–268.

Cascão, A. E., 2012, 'Nile water governance', in Awulachew, S. B., Smakhtin, V., Molden, D., and Peden, D. (eds.), *The Nile River Basin: Water, Agriculture, Governance and Livelihoods.* New York: Routledge: 229–252.

Cascão, A. E., and Zeitoun, M., 2010, 'Power, hegemony and critical hydropolitics', in Earle, A., Jägerskog, A., and Öjendal, J. (eds.), *Transboundary Water Management: Principles and Practice.* London: Earthscan: 27–42.

Chesworth, P., 1994, 'History of water use in the Sudan and Egypt', in Howel, P. P., and Allan, J. A. (eds.), *The Nile: Sharing a Scarce Resource: A Historical and Technical Review of Water Management and of Economical and Legal Issues.* Cambridge University Press.

Conway, D., 1996, 'The impacts of climate variability and future climate change in the Nile basin on water resources in Egypt', *International Journal of Water Resources Development*, 12(3): 277–296.

Conway, D., 2005, 'From headwater tributaries to international river: observing and adapting to climate variability and change in the Nile basin', *Global Environmental Change*, 15(2): 99–114.

de Waal, A., 1991, *Evil Days: Thirty Years of War and Famine in Ethiopia: An African Watch Report.* New York: Human Rights Watch.

de Waal, A., 1997, *Famine Crimes: Politics and the Disaster Relief Industry in Africa.* Bloomington: Indiana University Press.

Earle, Anton, Nordin, Kikki, Cascão, Ana Elisa, Rukundo, Drake, Seide, Wondwosen Michago, and Björklund, Gunilla, 2013, *Independent Evaluation of the Nile Basin Trust Fund (NBTF), Final Report*, Washington, DC: World Bank Group, http://documents.worldbank.org/curated/en/2013/06/19779850/independent-evaluation-nile-basin-trust-fund-nbtf-final-report

El-Raey, M., Dewidar, K. R., and El-Hattab, M., 1999, 'Adaptation to the impacts of sea level rise in Egypt', *Mitigation and Adaptation Strategies for Global Change*, 4(3–4): 343–361.

Elwan, N. M., and Hassaneen, H. M., 2002, 'Synthesis and reactions of indane-1, 3-dione-2-thiocarboxanilides with hydrazonoyl halides and active chloromethylene compounds', *Heteroatom Chemistry*, 13(7): 585–591.

ENTRO, 2007, *Eastern Nile Watershed Management Project*, Eastern Nile Technical Regional Office, Entebbe.

ENTRO, 2008, *Climate Smart/Proof ENSAP Projects*, Eastern Nile Technical Regional Office, Entebbe.

ENTRO, 2009, *ENTRO Approach Paper*, Eastern Nile Technical Regional Office.

ENTRO, 2011, *Nile-Flow: A Quarterly Newsletter of the Eastern Nile Technical Regional Office, September 2011.* Eastern Nile Technical Regional Office, Entebbe.

ENTRO, 2013, 'Baro-Akobo-Sobat Multipurpose Water Resources Study (BAS)', accessed 13 November 2014 at: http://entroportal.nilebasin.org/BAS/Pages/bas.aspx, Eastern Nile Technical Regional Office, Entebbe.

ENTRO, 2014, 'Flood Preparedness and Early Warning Project', accessed 13 November 2014 at: http://entroportal.nilebasin.org/FPEW/Pages/fpew.aspx

Ethiopia, 2011, *Ethiopia's Climate-Resilient Green Economy – Green Economy Strategy*, Federal Democratic Republic of Ethiopia.

Gill, P., 2010, *Famine and Foreigners: Ethiopia Since Live Aid*. Oxford: Oxford University Press.

Goulden, M., Conway, D., and Persechino, A., 2009, 'Adaptation to climate change in international river basins in Africa: A review', *Hydrological Sciences Journal*, 54(5): 805–828.

Handino, M. L., 2014, *Green famine in Ethiopia: understanding the causes of increasing vulnerability to food insecurity and policy responses in the Southern Ethiopian highlands*, doctoral dissertation, University of Sussex.

Heltberg, R., Siegel, P. B., and Jorgensen, S. L., 2009, 'Addressing human vulnerability to climate change: toward a "no-regrets" approach', *Global Environmental Change*, 19(1): 89–99.

Hillbruner, C., and Moloney, G., 2012, 'When early warning is not enough – lessons learned from the 2011 Somalia Famine', *Global Food Security*, 1(1): 20–28.

Howell, P. P., and Allan, J. A., (eds.), 1994, *The Nile: Sharing a Scarce Resource: A Historical and Technical Review of Water Management and of Economical and Legal Issues*. Cambridge: Cambridge University Press.

Hulme, M., 1994, *Global Climate Change and the Nile Basin: The Nile Sharing a Scarce Resource*. Cambridge: Cambridge University Press: 139–162.

IGAD, 2003, IGAD Climate Prediction and Applications Center, accessed 16 February 2015 at http://www.icpac.net/

Jarso, J. F., 2011, 'East African community and the climate change agenda: An inventory of the progress, hurdles, and prospects', *Sustainable Development Law & Policy*, 12(19).

Jeuland, M., 2010, 'Economic implications of climate change for infrastructure planning in transboundary water systems: an example from the Blue Nile', *Water Resources Research*, 46(11).

LVBC, 2003, Protocol Establishing LVBC, accessed 12 November 2014 at http://www.lvbcom.org/index.php/who-we-are/protocol-establishing-lvbc.

LVBC, 2013, Integrated Climate Change Adaptation and Mitigation Programme in the Lake Victoria Basin. Lake Victoria Basin Commission, Series No. 007, http://library.lvbcom.org:8080/handle/123456789/239

Maxwell, D., and Watkins, B., 2003, 'Humanitarian information systems and emergencies in the Greater Horn of Africa: logical components and logical linkages', *Disasters*, 27(1): 72–90.

MWRI, 2005, *National Water Resources Plan for Egypt 2017*, Cairo.

NBI, 2002, The Nile Basin Initiative Act.

NBI, 2012a, *Nile News*, 9(2).

NBI, 2012b, 'State of the River Nile Basin Report', Nile Basin Initiative, Entebbe.

NELSAP 2011, NELSAP Programme brief & NELSAP factsheet, Nile Basin Initiative, Entebbe.

Nicol, A., and Cascão, A. E., 2011, 'Against the flow–new power dynamics and upstream mobilisation in the Nile basin', *Review of African Political Economy*, 38(128): 317–325.

Nile Cooperative Framework Agreement, 2010, accessed 3 November 2012 at http://internationalwaterlaw.org/documents/regionaldocs/Nile_River_Basin_Cooperative_Framework_2010.pdf

Pankhurst, R., 1992, 'The history of deforestation and afforestation in Ethiopia prior to World War I', in *Proceedings of the Michigan State University Conference on Northeast Africa*, East Lansing: 275–286.

Piguet, F., and Bawtree, V., 2003, 'Food crisis in Ethiopia: Drought or poor economic policies?' *Review of African Political Economy*, 30(97): 485–489.

Prowse, M., and Snilstveit, B., 2010, 'Impact evaluation and interventions to address climate change: a scoping study', *Journal of Development Effectiveness*, 2(2): 228–262.

Sherif, M. M., and Singh, V. P., 1999, 'Effect of climate change on sea water intrusion in coastal aquifers', *Hydrological Processes*, 13(8): 1277–1287.

Tvedt, Terje, 2000, *The River Nile and its Economic, Political, Social and Cultural Role: An Annotated Bibliography*. Universitetet i Bergen, Senter for utviklingsstudier, Bergen, Norway.

UNEP, 2010 Africa Water Atlas. UNEP, Nairobi, Kenya, http://www.unep.org/publications/contents/pub_details_search.asp?ID=4165

Verdin, J., Funk, C., Senay, G., and Choularton, R., 2005, 'Climate science and famine early warning', *Philosophical Transactions of the Royal Society B: Biological Sciences*, 360(1463): 2155–2168.

Ververs, M. T., 2012, 'The East African food crisis: did regional early warning systems function?', *The Journal of Nutrition*, 142(1): 131–133.

Walker, P., 2013, *Famine Early Warning Systems: Victims and Destitution*. New York: Routledge.

Waterbury, J., 1979, *Hydropolitics of the Nile Valley*. Syracuse, NY: Syracuse University Press.

Waterbury, J., 2002, *The Nile Basin: National Determinants of Collective Action*. New Haven, CT: Yale University Press.

World Bank, 2012, 'Environmental and social management framework (ESMF) for the Nile Basin Climate Resilient Growth Program of the Nile Basin Initiative – A program proposed for funding under NBTF and CIWA', Washington DC: World Bank, accessed 16 February 2015 at http://www-wds.worldbank.org/external/default/WDSContentServer/WDSP/IB/2012/07/30/000386194_20120730022327/Rendered/PDF/E30760EA00P1300inal0AppraisalStage.pdf

World Commission on Dams, 2000, *Dams and Development: A New Framework for Decision Making*. London: Earthscan.

Zeitoun, M., and Warner, J., 2006, 'Hydro-hegemony-a framework for analysis of transboundary water conflicts', *Water Policy*, 8(5): 435–460.

7   The Orange-Senqu River basin

Introduction

The Orange-Senqu River has its headwaters in the Lesotho's Maluti mountains at around 3,300 metres above sea level from where it flows into South Africa. It joined by its largest tributary (the Vaal River), later forming the border with Namibia before flowing into the Atlantic Ocean, having covered around 2,300 kilometres (Earle et al. 2005). Ephemeral streams link southern Botswana to the basin. The basin is one of the most used in southern Africa and supplies most of the freshwater needs of the Johannesburg region, South Africa's industrial heartland. The river system is regulated by more than 31 major dams and is a highly complex and integrated water resource with numerous large inter- and intra-basin transfers (ORASECOM 2011). The world's largest international transfer of water is the Lesotho Highlands Water Project (LHWP), whereby water is transferred by gravity to South Africa, earning Lesotho royalty income. The basin is important both to the major cities in the two countries (for Johannesburg, the water allows it to keep expanding, and in Maseru the government benefits form the foreign income) as well as to the predominantly rural population of the basin, where it supports small-scale as well as large-scale commercial agriculture (Schuermans et al. 2004). Currently, there is a study underway about the feasibility of transferring water from Lesotho via South Africa to Botswana's capital city, Gaborone (Earle et al. 2005).

In common with the Nile River, the Orange-Senqu has a well-watered headwaters region and an extremely arid downstream area. Rainfall in the Lesotho Highlands is high, commonly between 1,600–1,800 millimetres per year, with evaporation rates commensurately low. The mid-reaches and the major tributary of the Vaal River receive between 400–800 millimetres a year, while the last-third of the river experiences rainfall below 200 millimetres a year, dropping to 45 millimetres at the river mouth at Oranjemund (UNDP-GEF 2008). The natural climate and resultant flow regime is highly variable over time, with a low conversion of rainfall to runoff, typically around 10 per cent. Drought periods and flood periods (extending over several years each) typically follow each other (Kistin & Ashton 2008). A major projected impact of climate change is to increase this natural variability (UNDP-GEF 2008; ORASECOM 2011).

In contrast with the Nile River, the Orange-Senqu waters are little-used in the downstream reaches, with low population densities in the arid desert zone. The vast majority (around 95 per cent) of the water is used upstream – by South Africa. Indeed, the industrialised area around Johannesburg (Gauteng province) is almost 100 per cent reliant on water from the basin (Heyns et al. 2008). This region generates 60 per cent of the South African GDP and supplies water to almost 20 million people (DWA 2012). Uses include domestic, electricity generation (cooling of thermal power stations), industrial (including petro-chemical), mining, and irrigation.

Total basin-wide water demand for 2010 is estimated at 5,867 million cubic metres, out of an annual flow of around 11,300 million cubic metres (UNDP-GEF 2008). The volumes of water (million cubic metres) used by the countries were 5,531 in South Africa, 134 in Namibia, and 23 in Lesotho. Botswana is deemed not to use any of the surface waters of the basin.

Although the amount of water used by Namibia is relatively small, it is nonetheless important for sustaining irrigated crop production (supporting a lucrative export fruit-trade), and mining operations along the lower reaches of the river. Irrigated water use in Namibia is expected to increase substantially over the coming decade, with a predicted total water use for Namibia in 2025 being 244 million cubic metres (UNDP-GEF 2008). As with Egypt in the Nile River basin, agriculture in Namibia on the banks of the Orange-Senqu River is wholly dependent on irrigation water, with no effective rainfall being received, making the river vital to the economic development of the country (Heyns et al. 2008). Water use in Lesotho is low and not expected to increase dramatically – reaching around 26 million cubic metres by 2025 (UNDP-GEF 2008). The urban development of Maseru has meant that the city has had to source water supply from ever further away, with the Metolong Dam now being constructed to supply the population.

The South African Department of Water Affairs anticipates that the South African population will grow at an annual rate of between 1.1–2.2 per cent until 2025, and expects water use will continue to rise three times faster than the rate of population increase (DWAF 2004). If this rate of water use increase materialises, it will place the Vaal River tributary in a water-deficit situation by 2025. The response has been to construct an intra-basin transfer scheme to convey water from the headwaters of the Orange-Senqu in the Lesotho Highlands to the upper-reaches of the Vaal River in South Africa. The first phase of the LHWP now supplies the Gauteng region with water, and construction of the second phase is now proceeding. The National Water Resources Strategy states that

> given the limited availability of fresh water resources, it is unlikely that it will be economically feasible to meet all the demands that may arise. Unless new innovative approaches are adopted in order to reconcile demand and supply, particularly in the most water-stressed catchments and areas of development, South Africa's growth will be negatively impacted.
>
> (DWA, 2012: s. iii)

In addition to water-quantity challenges, the country also faces a deterioration of water quality in several catchments, most notably the Vaal River, where salinity levels are high. Agricultural return flows, municipal releases, and acid mine drainage are all contributing to a reduction of water quality in this stretch of the river. This has necessitated the release of extra water from upstream in an effort to dilute the pollution levels in the Vaal River (DWA 2012: s. 212).

As mentioned, there is the possibility of transferring water from Lesotho, via South Africa, to supply the capital city of Botswana, Gaborone. The volume of water consumed is not likely to be large, however the impact may be important as this will involve the direct transfer of water to the Limpopo River basin, depriving the Orange-Senqu of return-flows.

South Africa exercises strong hegemonic power in the basin, with a population and economy an order of magnitude larger than the other basin states combined. The economies of the other three basin states are highly reliant on and highly integrated in the South African economy with significant flows of migrants and goods. The currencies of Namibia and Lesotho are directly coupled with the South African rand; and the Botswana Pula trades at a rate determined through a basket of currencies which includes the rand. Despite the disparities in power and capacity among the basin states, the relations between them are good, with cooperation proceeding in a variety of sectors, including electricity supply, development of transport routes, law enforcement, and water.

Multilateral approaches to managing the basin only started to emerge after the advent of democracy in South Africa in 1994. Prior to that, most of the cooperation was bilateral, mainly involving South Africa and another basin state. Various management organisations were formed for implementing bilateral projects including the Lesotho Highlands Water Commission (LHWC) to implement the Lesotho Highlands Water Projects between South Africa and Lesotho; and the Permanent Water Commission between South Africa and Namibia to implement the Vioolsdrift-Noordoewer Joint Irrigation Scheme (VNJIS). In 2000, the four basin states negotiated the formation of the Orange-Senqu River Commission (ORASECOM), constituting it as an international organisation to advise the parties on the equitable and sustainable development of the basin (Earle et al. 2005).

The commission provides a forum for consultation and coordination between the basin states and is currently in the process of adopting a basin development plan (NBI 2011). ORASECOM advises the member states on matters related to development, use and conservation of the water resources in the river system. This may include recommendations on: water availability, equitable and reasonable use of water resources, development of the river system, stakeholder participation, and harmonisation of policies (ORASECOM 2000).

The commission is controlled by the Council – where the four governments are represented by three Commissioners each. At present, all the Commissioners are from government departments (mainly the respective ones responsible for water issues), though nothing in the formation agreement precludes the appointment of non-governmental representatives as Commissioners

(ORASECOM 2000; NBI 2011). The Council may mandate the formation of technical task teams – to perform studies on specific issues. There is a standing technical task team – responsible for a range of hydrological and related studies – as well as a legal task team and a communications task team. The daily functioning of the Commission is supported by a secretariat, consisting of four recruited staff (NBI 2011).

GCC entering the basin

Projected impacts of GCC

The Orange-Senqu River basin is one of the most developed and most utilised river basins in southern Africa and possibly on the continent of Africa. As discussed above, it is a key part of South Africa's water security, as well as being an important asset to the other three basin states. There is very little excess or spare water remaining in the river, a situation compounded when environmental considerations are taken into account – something not adequately done today. Consequently, any change in the availability of water in the basin due to climate change is of great interest and importance to the basin states and other actors. A consequence of the high level of development of the basin's water resources is that the system is extensively measured and modelled, and knowledge about natural climatic variability is high. A number of studies have sought to understand and project the possible impacts of climate change on the basin. Though there is some considerable variation in the detail of these studies, they all broadly point in one direction – under projected global climate change scenarios the Orange-Senqu River will become more like itself.

Models predict that as temperatures across the basin increase there will be a drop in rainfall in the western portion of the basin (the part which is very dry today) and an increase in rainfall in the eastern part, specifically over the Lesotho Highlands (Kistin & Ashton 2008; Knoesen et al. 2009; Kranz et al. 2010; ORASECOM 2011). Inter-seasonal variability is set to increase; and the range of high and low flows will increase, meaning that flooding and droughts will become more pronounced.

The European Union-funded NeWater project sought to develop adaptive responses to the challenges being faced in various transboundary basins, including the Orange-Senqu. As part of this project, a climate change scenario for the basin was developed, seeking to describe the impacts of a change in temperature and rainfall on runoff in the basin (Knoesen et al. 2009). The study was performed by a team from the University of Kwa-Zulu-Natal and based on long-term research initiatives around the climatology and hydrology of the basin. Runoff for 1,443 quaternary catchments (fifth-level sub-basins) was calculated using the ACRU model (a model which has been applied extensively across southern Africa). Inputs to the model included daily rainfall, temperature, and evaporation data as well as information on soils and vegetation cover for each of the sub-basins. A moderate scenario for the future climate across southern

Africa was chosen – coded ECHAM5/MPI-OM. This Global Circulation Model (GCM) uses the 'A2 emissions scenario, which assumes that GHG emissions continue relatively unabated to the year 2100' (Knoesen et al. 2009: s. 5). Based on this data, hydrological scenarios were developed for three time periods: the present (1971–1990), the intermediate future (2046–2065), and the more distant future (2081–2100).

The results of this model indicate an increase in temperature across the basin for the intermediate future, ranging from 1.5°C in the lower basin to 2.0°C in the Lesotho Highlands (Table 7.1). Likewise evaporation (which includes evapotranspiration in the NeWater study) increases across the basin, with a greater increase in the Highlands. Rainfall is projected to increase across the basin by around 20 per cent, with a greater increase in the Highlands accompanied by a small decrease in the lower basin. Changes in precipitation are modified by evaporation, soil type, and ground cover to produce an impact on streamflow, often amplifying the impact up or down. For this reason, there is considerable local variation across the basin in streamflow, but with a general pattern of an increase in the Highlands and a drop in the lower basin.

According to the study the impact of these changes is that flooding is likely to become more common in the basin, with a potential for causing harm to human settlements and activities, such as farming. Generally it is found that droughts (especially short-duration ones) will become less frequent in the east of the basin and more frequent in the west (Knoesen et al. 2009: s. 28).

A more recent analysis on the impacts of global climate change for the basin was performed under the ORASECOM Basinwide IWRM planning study. A projection of impacts under plausible scenarios and guidelines on climate changes adaptation strategies was produced by WRP Consulting Engineers and other associates (ORASECOM 2011). This study also used the ECHAM5 GCM (running the A1B emissions scenario), but complements this with projections made by a statistical model called STAR II, based on historical weather records available for the region. Impacts of climate change were modelled for temperature as well as precipitation for the basin for the period 2051–2060 (see Table 7.1).

Table 7.1 Modelled impacts of climate change under NeWater and ORASECOM studies

Parameter	*NeWater (2046–2065)*		*ORASECOM (2051–2060) – median realisation*	
	Lower basin	*Highlands*	*Lower basin*	*Highlands*
Temperature	+1.5 – 2.0°C	+2.0 – 2.5°C	+1.0°C	+1.5 – 2.0°C
Evaporation	+10%	+10 – 15%		
Precipitation	−10%	+20 – 30%	−40 – 60mm	+20 – 40mm
Stream flow	−20%	+30 – 100%		

(Source: Knoesen et al. 2009 and ORASECOM 2011)

The ORASECOM study found that temperature increase in the basin would be similar to that proposed by the NeWater study – varying between 1 degree Celsius at the mouth and 1.5 to 2 degrees in the Highlands. Three 'realisations' for precipitation were produced – dry, median, and wet. There is some considerable difference between the results of the three, with the dry realisation showing a drop of 140mm a year in the northeast of the basin, while the wet realisation shows a maximum drop of around 80mm and an increase in rainfall over the Highlands of 20–60mm. In the median realisation, the increase in the Highlands is between 20mm–40mm, while the average for the basin is a 60mm drop in rainfall (ORASECOM 2011). Note that the NeWater study presented changes in precipitation as a percentage of the current value while the ORASECOM study presents quantities. If a maximum current annual precipitation figure for the Highlands is taken as 1,800mm, a 20–30 per cent increase (as projected by NeWater) would equate to a 360–540mm increase, an order of magnitude greater than the median realisation proposed by ORASECOM. The researchers in the ORASECOM study found that the modelled precipitation figures produced using the GCM 'proved unsatisfactory' and relied on the statistical model for these. This points to some of the challenges in downscaling global models to regional or basin level, making planning and adaptation to the impacts of climate change more difficult.

The ORASECOM study could not identify a change to the current climate, pointing out that the high level of natural climatic variability makes it difficult to detect more-subtle long-term changes in climate (ORASECOM 2011:s. 12). The basin experiences periods of high flows and periods of low flows (compared with the long-term average), with these cycles taking up to 20 years.

It is not the intention of this publication to debate the facts and predictions around climate change, with the interest being to analyse the responses at a transboundary level to the climate change debate. What can be concluded, though, from the two studies above is that the basin overall may become somewhat drier, but with a difference between the eastern Highlands region, which may become wetter and the western lower basin which will become markedly drier than today. This precipitation is also likely to be received through short-duration rainfall events, leading to possible problems such as flooding, increased soil erosion, and sedimentation, and reduced groundwater recharge (ORASECOM 2011: s. 17). It is possible that the increase in runoff from the Highlands will more than offset the decrease in runoff from the lower reaches, introducing the possibility of more water being available to the system overall. This water would only be usable if the infrastructure exists to make use of it and it is possible that current dams would not be large enough to effectively capture the higher flood peaks. There would also be a negative impact on rainfed agriculture across the basin, an important factor to consider given that this compromises a large portion of agricultural production, especially when areas relying only on supplementary irrigation are taken into account. In essence, reliance on perennial irrigation would increase. Both studies point out that there is still a high degree of uncertainly around the predicted changes in precipitation, and likewise the commensurate changes in runoff. A key recommendation is to improve data collection and monitoring (of rainfall, temperature, and stream-flow) across the basin.

Perceived impacts of GCC

Generally within the popular media amongst the basin countries, climate change as an issue receives a lot of attention. Statements about the severity of the problem are frequently made in public in a variety of sectors, with a general support for the findings of the WWF in South Africa which proposed, based on extensive studies on a range of impacts of future climate change, 'It's nearly impossible to overstate the threat of climate change' (WWF-SA 2013). If the real impacts of climate change (as distinct from natural climate variability) have not yet been felt in the basin, the perceived impacts of climate change (or global warming, as it is still popularly referred to in the region) are certainly receiving attention at various levels in all four basin states.

The expectation that climate change will have an impact is brought up in conversation in the region, such as the relatively mild winter of 2013 being ascribed to the results of climate change – as opposed to natural climatic variability. In parallel with the popular understanding of the impact of climate change which is frequently not based on scientific evidence, there is a trend of basing media reports about climate change on science. Recent newspaper articles (from the past five years) make an effort to go beyond the rhetoric and to understand some of the complexities and nuances. For instance, an article in the *Mail and Guardian* (a weekly 'serious' newspaper in South Africa) delineates the interconnected nature of water scarcity and energy scarcity in the region. Since 2005, the southern Africa region has been faced with constraints in the supply of electricity, resulting in intermittent power outages. The undersupply situation is largely the result of higher-than-anticipated economic growth in the decade after the dawn of democracy in South Africa in 1994, and the success of connecting previously un-served sections of the population to the grid in that country. Due to the inter-connected nature of the power grids of the four basin states, there has been a knock-on impact in each.

The article shows that although the electricity scarcity in the region is a serious concern, it is ultimately solvable through the construction of new power generation facilities – something happening at present in three of the basin states. However, it points out that these power generation facilities – either coal-fired in the case of South Africa and Botswana or hydro-powered in Lesotho – all rely extensively on the availability of water resources. The statistic presented in the article is that a kilowatt-hour of electricity supply requires a kilogram of coal and two kilograms of water to produce. In addition, the mining and the burning of coal has negative impacts on water quality across the region, leading to acidification of watercourses (Mail & Guardian 2011). The article points out that South Africa is one of the largest emitters of greenhouse gases in the world (per person) and establishes the link between these emissions and a future of more-constrained water supplies due to the impacts of climate change. Interestingly, the article proposes that South Africa will have to rely increasingly on transboundary water transfers from neighbouring states, specifically mentioning the existing flow from Lesotho.

In Lesotho, there is a large amount of reporting on climate change in the media, with the central theme being increased variability – whether causing unseasonably cold weather or unseasonably dry weather, or high rainfall or low rainfall. A theme which emerges is that citizens, urban as well as rural, need to be informed on issues related to climate change, as it will have a direct impact on their ability to make a living (Public Eye 2013). This sentiment is echoed in Botswana, where several media reports stress the need for public education on the impacts of climate change, most specifically on the agriculture sector (*Daily News Botswana* 2013). For this to be effective, one of the articles concludes (see *Daily News Botswana* 2013), there needs to be better long-term forecasting of weather conditions by the meteorological services of the country. In Namibia, in addition to the issue of food security, there is also emphasis on the threats to the country's bio-diversity from a long-term change in climate. The country relies extensively on the tourism sector to stimulate economic development and create jobs, and much of this tourism is explicitly linked to the high biodiversity in the country as well as the large number of endemic plant species. Newspaper articles such as that in *The Namibian* report a firm link between climate change and threats to the country's natural biodiversity (*The Namibian* 2013).

Looking at a selection of media (by no means a fully representative sample from the four countries, rather just a short overview for writing this chapter), it emerges that the bulk of the articles around climate change in the four basin states generally – and in the Orange-Senqu basin specifically – deal with the perceived or predicted impacts of climate change. Principally, this boils down to an increase in temperature across the basin, an increase in climatic variability across the basin (more droughts, more floods), and some acknowledgement that there is likely to be a long-term general drop in precipitation in the four basin states but with a possibility of increases in some areas – such as the Lesotho Highlands and eastern areas of South Africa. When it comes to reports on current affairs linked with climate change in the region, most are about flooding. In all four basin states, the decade leading up to 2013 has been wetter than usual, with flooding in the mid to lower reaches of the Orange-Senqu River becoming a common phenomenon. In a newspaper article on coastal flooding in southern Africa as a result of climate change, the author also links flooding due to increased rainfall over inland areas, with the caption accompanying a photo of a farm homestead surrounded by water, stating, 'Floodwaters from the swollen Orange River devastated the Northern Cape early this year. Even inland regions are not immune from the effects of climate change' (IOL 2011). Flooding seems to attract the bulk of the headlines in the media and has certainly been an issue in the basin states over the past decade. This could also be due to the more-visible nature of flooding – graphic images of roads and bridges washed away make for more compelling news television pictures than maize plants whose growth has been stunted due to a lack of water.

On a national level the four basin countries have taken concerted steps to develop preparedness for the impact of climate change on water resources. A brief snapshot of national initiatives and responses is provided below; as the focus of this publication is on the transboundary responses to the climate change debate these national responses are not explored in detail.

Botswana

Botswana signed and ratified the Kyoto Protocol in 1994, but has not developed a national climate change policy since then. This is in the process of being corrected through an initiative to develop a National Climate Change Policy (the Policy) and Comprehensive National Climate Change Strategy and Action Plan. The Policy will incorporate a climate change situation analysis which identifies the country's primary vulnerabilities and how these will impact economic sectors. In addition the contribution of the country's economic sectors to GHG emissions will be reviewed. In response a policy and institutional framework will be developed to enhance implementation of climate change mitigation as well as adaptation in the country.

Following from the Policy a National Climate Change Strategy and Action Plan is being developed. This will consist of the following components:

- Long-term low-carbon development pathway/strategy – seeking to shift towards more efficient use of natural resources, including water,
- National Adaptation Plan (NAP) – identifying priority actions for promoting resilience to climate change impacts on a national level but also on the transboundary level,
- Nationally Appropriate Mitigation Actions (NAMAs) – identifying actions which could be taken and funding streams which can support national actions to mitigate GCC,
- Technology development and transfer – focusing on developing local skills and capacity for research and manufacturing,
- Strategy for communication, knowledge management, and capacity development – which will aim to reach out to all sectors and stakeholders; and
- Financial mechanisms – identifying models for establishing climate resilience funds to support investments in promoting resilience to GCC in the country including through the development of water resources infrastructure.

The National IWRM and Water Efficiency plan is set to be concluded by the end of 2013. This plan identifies GCC as a major risk for water resources management in the country. It emphasises the need to build mechanisms for adapting to a changed climate through measures such as water use efficiency and wastewater reuse, as well as investigating the feasibility of inter-basin transfers.

Lesotho

As a least-developed country, Lesotho has developed a National Adaptation Programme of Action (NAPA). This NAPA was developed with donor support – from the UNFCCC process as well as financing from the Global Environmental Facility (GEF). The key objectives of the NAPA process entail: identification of communities and livelihoods most vulnerable to climate change, generating a list of activities that would form a core of the national adaptation programme of action, and communicating the country's immediate and urgent needs and priorities for building capacity for adaptation to climate change. The NAPA identifies a range of negative impacts on various communities and economic sectors in the country and responds with 11 priority projects to build resilience. Several of these involve improved use of water resources, through actions such as wetland conservation and minimisation of soil erosion.

The objectives of the Lesotho National Strategic Development Plan (NSDP) 2012–2016 require development goals to be delivered in a climate-resilient manner. Taking action in relation to environment and climate change is one of its five strategic axes. Concerns with regard to climate change notably relate to its potential impacts on agriculture, water availability, soil erosion, mountain livelihoods, biodiversity, and disaster risks.

Namibia

The country has developed a National Strategy on Climate change, dealing both with adaptation as well as mitigation issues. The four thematic focal areas for climate change adaptation efforts are:

- Food security and sustainable resource base,
- Sustainable water resources management,
- Human health and well-being, and
- Infrastructure

Water is obviously a cross-cutting domain under all four themes and is referenced under all of them. Under the sustainable water resources management theme, the emphasis is placed on improving watershed conservation and management as well as promoting an integrated approach to planning.

South Africa

A National Climate Change Response Strategy has been developed and is in the process of being implemented and develops risk-based processes to identify and prioritise short- and medium-term adaptation interventions to be addressed in sector plans. The processes will also identify the adaptation responses that require coordination between sectors and departments and it will be reviewed every five years. For the immediate future, sectors that need particular attention are water, agriculture and forestry, health, biodiversity, and human settlements. Resilience to climate variability and climate change-related extreme weather events will be the basis for South Africa's future approach to disaster management and we will use region-wide approaches where appropriate.

The strategy recognises that while there is a degree of uncertainty as to the net effects of climate change on water availability, rainfall is expected to become more variable, with an increase of extreme events, such as flooding and droughts, resulting in a much more variable runoff regime. Downscaled climate modelling suggests that the western and interior parts of the country are likely to become drier, and the eastern parts of the country wetter. Increased rainfall intensity will exacerbate scouring in rivers and sedimentation in dams, potentially impacting on water supply and treatment infrastructure. Higher temperatures, combined with higher carbon dioxide levels, will contribute to increased growth of algae as well as faster evaporation rates negatively impacting water resources.

Basin-wide response

Organisation formation

As introduced previously, the four basin states are advised on issues related to the sustainable management and development of the Orange-Senqu basin by ORASECOM, the basin-wide RBO. Specifically in Article Five of the ORASECOM agreement, the organisation is mandated by the parties to take all measures required to make recommendations, or to advise the parties, on the following matters (citing a selection from Article Five of the ORASECOM formation agreement):

- 5.2.1: Measures and arrangements to determine the long-term safe yield of the water sources in the River System;
- 5.2.2: The equitable and reasonable utilisation of the water sources in the River System to support sustainable development in the territory of each party;
- 5.2.5: The standardised form of collecting, processing, and disseminating data or information with regard to all aspects of the River System.

(ORASECOM 2000)

The activities mandated above potentially open a large opportunity to develop tangible responses to climate change in the basin along the climate change adaptation chain introduced in Chapter One of this volume. Recall that these adaptation actions can be at the organisational level, where staff with the required skills are employed (such as climate change specialists or flood and drought forecasters) or when training on these technical issues is promoted; at the systems level, by setting in place early-warning systems or dam synchronisation mechanisms; or at the structural level, by increasing the storage capacity of dams or developing inter-basin transfers or building flood defences, in order to deal with climate change. Due to the natural variability in the basin climate (droughts and floods have always been present), it is officially recognised that most water-users and other stakeholders (including the environment) currently possess a relatively high degree of resilience to the impacts of climate change (ORASECOM 2011). Adapting to changes in temperature and precipitation between one year and another is the norm. The question is, to what degree are the current institutional and management mechanisms in the basin able to cope with an increase in climatic variability; and what actions should be taken at the basin-wide scale to increase resilience?

Due to the basin lying in a semi-arid to hyper-arid zone, coupled with the different levels of economic development in various parts of the basin, the emphasis of multilateral management approaches (whether involving all the basin states or only a selection) has been on intra- and inter-basin transfer schemes. This, along with the need to secure environmental flow requirements, is likely to remain the focus for future basin-wide management interventions (Kistin & Ashton 2008). These transfers deserve some deeper analysis, as they have the greatest potential to impact other water-users and the environment in the basin and in turn could be impacted by changes in the natural climate. Most of the agreements to pursue water transfers schemes have been bilateral and pre-date the formation of ORASECOM.

According to the ORASECOM agreement, the institutions created for the implementation of the bilateral projects do not fall under the control of ORASECOM – they only have an obligation to liaise with it in accordance with the ORASECOM agreement (ORASECOM 2000).

To support the continued high rate of economic growth and population increase in the Gauteng region, it is likely that water will continue to be sourced from ever further away, with the Orange-Senqu headwaters in Lesotho being the most accessible source. The city of Johannesburg was founded in the late 19th century to exploit the vast gold reserves lying underground in that region. It is one of the largest cities in the world not located close to a perennial watercourse, the closest being the Vaal River (a major tributary of the Orange-Senqu), roughly 65 kilometres away (Earle, Malzbender, Turton, & Manzungu 2005). Additionally, the city straddles the watershed line between two river basins – the Orange-Senqu and the Limpopo. All water consumed in the northern suburbs of the city (as well as other urban areasm such as Pretoria which draws on the same water supply) flows north into the Limpopo River and is liberated in the warm Indian Ocean in Mozambique. Water used in the south of the city runs back into the Vaal River, eventually joining the main Orange-Senqu River and flowing into the cold waters of the Atlantic Ocean.

Infrastructure development

By the 1970s, the area around Johannesburg was facing severe water scarcity. The Lesotho Highlands option had been identified as a water supply for Johannesburg in the 1950s and discussions between the two countries proceeded through the

1970s (Robbroeck 2007). A joint feasibility study had described in some detail how the upper reaches of the Orange-Senqu River in the Lesotho Highlands could be dammed and the water transferred through a tunnel under the Maluti Mountains to South Africa (Earle, Malzbender, Turton, & Manzungu 2005). However, due to the political climate of the time, Lesotho did not want to enter into an agreement with South Africa. High-level political and technical delegations from the two countries met regularly, with South Africa giving the project great political support. Only when a new government came to power in Lesotho in 1986 was an agreement between the two countries reached, marking the start of the Lesotho Highlands Water Project (LHWP). This project formed a part of what was referred to as a 'total water strategy' for South Africa, linking the need for developing the project to national water security (Turton, Meissner, Mampane, & Seremo 2004).

Phase 1 of the project is now complete and supplies 770 million cubic metres of water a year to South Africa, earning Lesotho around 66 million US dollars in royalty fees during 2010 (LHWP 2011). In 2011, the two countries agreed to proceed with the second phase of the project, which will eventually increase the amount of water transferred to South Africa to 2,000 million cubic metres. In addition to the benefit of increased water flows, South Africa also enjoys more efficient storage of water resources, due to the lower evaporation rates in the Lesotho Highlands compared to dam locations close to Johannesburg (Earle, Malzbender, Turton, & Manzungu 2005). As a direct result of the LHWP, a bi-national management organisation was created by the two countries – the Lesotho Highlands Water Commission (LHWC) (originally called the Joint Permanent Technical Commission) tasked with overall oversight of the project (Earle, Malzbender, Turton, & Manzungu 2005). The World Bank at that stage placed pressure on South Africa and Lesotho to gain a statement of 'no objection' from downstream Namibia for the construction of the LHWP.

Due to the concerns of the downstream basin state, Namibia, about possible negative impacts of the LHWP, a process was started in 1995 to establish a basin-wide organisation (Kranz & Vidaurre 2008). This culminated in the formation of the Orange-Senqu River Commission (ORASECOM) in 2000, incorporating all four basin states including Botswana. The existence of ORASECOM has meant that the hegemonic dominance of South Africa over the other basin states is now challenged, providing the opportunity for integrated management and development of the basin (Ibid).

Interesting to note is that the LHWP (Phase I and Phase II) is not explicitly linked to providing security of supply in the face of climate change. The wording in official documents has much more to do with promoting the efficiency of the scheme – lower evaporative losses and lower cost compared with pumping the water to Johannesburg from the downstream part of the Orange-Senqu basin. There has been some degree of controversy in the region about proceeding with Phase II of the LHWP, with some commentators questioning whether it is worth the expense (estimated at over 1 billion USD), compared with the, potentially, cheaper option of implementing comprehensive water demand management strategies in the municipalities in Gauteng which receive water from the LHWP Phase I.

Lori Pottinger of the International Rivers Network (an anti-dam NGO) proposes that water conservation, reuse, and recycling actions would be more resilient to climate change impacts and would result in a lower impact on the environment (*Business Day* 2011). In support of Phase II, the former Director General of the South African Department of Water Affairs, Mike Muller, makes the case that 'what the LHWP uses is Lesotho's land and gravity, to bring water to Gauteng and surrounds, which is much more efficient than pumping it 400m uphill from Aliwal North into the Vaal River. That would use a power station and a lot of electricity – and, in an era of climate change, energy saving is vital. LHWP may even generate some clean hydropower, as well as deliver water. We owe a debt of gratitude to the engineers who had this brilliant idea in the fifties, before anyone even dreamt of climate change' (IMD 2013). Hence, the project is linked both to climate change adaptation as well as to climate change mitigation – important in South Africa, as the large majority of its electricity supply is derived from coal-fired thermal power stations with their attendant high GHG emissions.

The overall impact of the LHWP in the lower Orange-Senqu River will remain relatively benign as most of the water transferred to the Gauteng region finds its way back to the river as return flows through the Vaal tributary. However, a sizable portion crosses the watershed into the Limpopo basin, constituting an out-of-basin transfer. As the northern reaches of Johannesburg continue growing faster than the southern areas, this net transfer to the Limpopo may increase. Additionally, Lesotho, South Africa, and Botswana recently entered into a Memorandum of Understanding (MoU) to explore the possibility of transferring water from the Lesotho Highlands via South Africa to Botswana to supply the city of Gaborone with water. This would constitute a direct out-of-basin transfer, again into the Limpopo. Few details of this possible inter-basin transfer scheme are available, and at the time of writing, a study to investigate the feasibility of such a transfer was being initiated.

The other joint infrastructure project in the basin is at a much smaller scale and is operated by Namibia and South Africa. The Vioolsdrift-Noordoewer Joint Irrigation Scheme (VNJIS) was initiated in 1987 (prior to Namibian independence in 1990) and formalised through a treaty in 1992. Under this treaty, the irrigation scheme allocates 9 million cubic metres to farmers in Namibia and 11 million cubic metres to farmers in South Africa annually (VNJIS 1992). No water is transferred out of the basin and the only losses are due to evapo-transpiration.

Future developments

An important factor to consider in the climate change resilience of water management and development schemes is the degree of flexibility built into the foundation agreement; ideally, these should incorporate mechanisms whereby the parties can adapt to short-term as well as long-term changes in climate and other factors (Kistin & Ashton 2008). In a comprehensive study on treaty flexibility in the Orange-Senqu river basin, Kistin and Ashton investigate five sets of mechanisms for enhancing treaty flexibility in the basin: allocation strategies, drought response provisions, amendment and review processes, revocation clauses, and institutional responsibilities (Kistin & Ashton 2008: s. 390).

According to their analysis, the LHWP treaty (with its six protocols) possesses attributes preparing it to respond to climatic changes across all five sets of mechanisms. There is a clear allocation schedule for water deliveries to South Africa, which can be reviewed by the parties to respond to demand conditions in South Africa. During times of drought, Lesotho may reduce the amount of water delivered to South Africa, and then make up this shortfall over the successive six months. This allows for response to short-term large variations in climate, however it may be challenged during a long-term change in climate. Thus far, this mechanism has not been invoked.

The original LHWP agreement has three mechanisms allowing for amendment or review: a requirement for the parties to review the agreement every 12 years; a phased implementation period (Phase I is operational, Phase II construction proceeding and phases 3 to 5 identified in the treaty) allowing parties to pull out of further phases (with financial penalties); and the possibility of amending the original treaty through rules to respond to changed or unforeseen circumstances (Kistin & Ashton 2008: s. 394). These rules could possibly be used to respond to long-term changes of climate in the basin, such as changes to the delivery schedule or amendments to the design, construction, and operation of infrastructure. To date, the only such amendment has been to the institutional framework established to oversee the project, changing from the construction phase to the operational phase. The feasibility study for Phase II of the LHWP proposed several changes to the design of infrastructure, compared with what was envisaged under the 1986 treaty. This related mainly to an alternative siting of one of the new dams (Polihali dam being recommended over Mashai dam). This, however, had nothing to do with climatic or hydrological conditions in the basin and was driven by the fact that the Mashai dam would have entailed pumping water to Katse dam (before gravity flow to South Africa), whereas transfers from Polihali dam would be by gravity flow only. This saves on electricity costs, something which was less of an issue at the time of signing the 1986 agreement, but which is now a problem in the region (LHWC 2009).

The agreement also has a revocation clause – though possibly incurring financial penalties for the party which initiates it. Finally, an institutional mechanism is created through the Lesotho Highlands Water Commission (LHWC), a bilateral body which oversees the construction and operation of the project.

It is possible to conclude that for the single largest and most important multilateral project in the basin, there is the possibility to respond to global climate change. To date, there has been no formal discussion of this, which is an interesting point given that studies show that there is likely to be an increase in precipitation over the Highlands region. If there were a long-term increase in rainfall over the Highlands, it would open up several other possibilities. First would be that environmental water requirements downstream of the LHWP could be met more easily. Currently, this is not being done, with several reaches of river experiencing water quality problems as a result and the wetlands of the delta area reducing in size. If more water were allowed to flow naturally out of Lesotho, it would be easier to satisfy the environmental requirements.

A second option would be to increase the amount of water transferred through future phases of the LHWP to South Africa. This is unlikely to be needed before 2055, according to projections prepared for the feasibility study of Phase II, but in a situation of long-term water scarcity around the Gauteng region could provide a useful extra supply (LHWC 2009). There is also the possibility to allow more Lesotho Highlands water to flow down the Vaal River, improving water quality through dilution, a feasible yet costly option to the water quality problems in that part of the river (UNDP–GEF 2008).

The third option is to increase the amount of water delivered to South Africa and then to transfer this to Botswana. This may be needed under climate change scenarios, many of which indicate most of Botswana to become considerably drier. The MoU between Botswana, Lesotho, and South Africa was signed by the respective ministers responsible for water affairs from each country on 1 March 2013 in Gaborone, with full reference to ORASECOM and in the presence of the Namibian minister for water affairs. Though the supply of water to Botswana from the Lesotho Highlands may build resilience to the impacts of a drier climate, there is no mention to this in any of the official documents around the initiative. Mention is made of Botswana's natural state of water scarcity, now placed under greater stress due to economic and population growth in the country.

A fourth option for how to treat any long-term increase in rainfall exists. That would relate to how the water for Phase II of the LHWP is transferred to South Africa. Currently the water is scheduled to be delivered to the country according to its needs (aimed at maximising assurance of supply); however, there is a possibility to transfer water directly from the new dam (Polihali) to the existing Muela hydropower station before being released to South Africa. This option was explored as part of the negotiations between South Africa and Lesotho on Phase II of the Highlands scheme, and was the one favoured by Lesotho as it would have maximised the amount of electricity it could produce (Quibell 2013). This option was expected to result in increased assurance of supply risks in South Africa, and an economic analysis suggested that the potential losses to South Africa would be far greater than the benefits to Lesotho. The option was therefore not considered further, and the possibility of pumped storage hydropower at the Katse Dam was offered as an alternative.

Responses in/from the basin – adapt, resist, subvert

Adapt

At the organisational level, the basin-wide commission, ORASECOM, has within its mandate the responsibility to advise the parties on future development issues in the basin – including responding to emergent issues, of which climate change could be one. On the basin–wide scale, there have been several initiatives to study the impacts of climate change on the basin, as discussed earlier. As ORASeCOM has a fairly well-developed organisational framework in place, it should in theory prove possible to develop systems for the optimisation of current infrastructure as well as management interventions to enhance adaptation capacity. However, in practice, this organisational capacity has not resulted in significant changes to the way the basin is managed. For instance, a water resources yield model has been developed and agreed on for the basin. This is an essential step towards better managing and developing the basin and could lend itself to building resilience to climate change. However, the model does not incorporate any climate change scenario – relying instead on the existing climatic variability parameters. This approach generally works well; incorporating scenarios which include a successive number of driest years or a successive number of wettest years and calculating the basin-wide yield on that. However, a sustained trend of dry years getting drier or wet years getting wetter would need adjustment of the yield model (Quibell 2013). As has been found above, the development of new infrastructure in the basin is not being explicitly tied to a climate change response, with issues of water security being at the fore.

Resist

A basin-wide perspective, considering the needs of all the basin states and the possible impacts of climate change (increased rainfall in the Highlands), may well have resulted in a different perception of the risks and benefits of Lesotho's preferred option of maximising hydropower generation under Phase II of the LHWP. What emerges in the Orange-Senqu basin is that multi-lateral activity is driven very much by the needs and interest of individual states, with South Africa being the most powerful and dominant. The resulting infrastructure initiatives have some degree of flexibility built into them, allowing them to respond to future changes in climate, but these mechanisms have not yet been used in response to changed climatic or hydrological conditions. In effect, the possibility of a changed future climate seems to be ignored at the transboundary level. Not in the sense of including the terms 'climate change' and 'adaptation' or 'mitigation' in most reports and other documents prepared by ORASECOM and other actors at the transboundary level, but certainly in the sense that nothing concrete is done to prepare for the possible effects of climate change. The prospect of increased precipitation in such a generally arid part of the world is surely something of which to take note. It was not possible during the research for this chapter to find a suitable explanation for this resistance for taking concrete action to plan for the eventuality of increased precipitation. It could possibly be that there is still not enough confidence in the predictions of increased rainfall – understandable, given the large degree of variation between studies.

However, if water scarcity is such a concern in the region, one may have expected at least some reconnaissance-level studies to investigate possible opportunities which may open up due to an increase in precipitation over the

Highlands. In effect, the tri-partite MoU between Botswana, Lesotho, and South Africa amounts to this – but not explicitly. It may also be that any increase in runoff generated in the Highlands would allow South Africa to ensure that it continues allowing an equitable share of the river to reach downstream Namibia – as well as the environmental requirements. This would allow South Africa to continue to increase its use of water upstream, or at the very least not need to take the politically difficult step of curtailing water-use, while serendipitously keeping its downstream neighbour happy.

Subvert

Looking at the way that a comprehensive approach to responding to climate change is lacking in the basin, it is difficult to find explicit instances of the 'subvert' reaction to the climate change debate. This may have to do with the fact that most of the basin states (Botswana, Namibia, and South Africa) are considered mid-level developing countries and are hence not eligible for much donor financing. Certainly bilateral donors have all but withdrawn from the three countries, with only limited multilateral or regional funding still being prevalent. Hence, there has been little subversion of existing agendas to fit the climate change debate with the expectation of accessing funding streams ordinarily not available to the countries.

References

Business Day, 4 October 2011, 'Colin Hoag and Lori Pottinger: Lesotho water project needs scrutiny', accessed August 2013 at http://www.bdlive.co.za/articles/2011/10/04/colin-hoag-and-lori-pottinger-lesotho-water-project-needs-scrutiny

Daily News Botswana, 26 April 2013, 'Metereological Services should make alerts', accessed August 2013 at http://www.dailynews.gov.bw/news-details.php?nid=2527

DWA, 2012, *National Water Resources Strategy 2,* Pretoria: Department of Water Affairs.

DWAF, 2004, *Orange River System: Overarching Internal Strategic Perspective.* Pretoria: Department of Water Affairs.

Earle, A., Malzbender, D. B., Turton A. R., and Manzungu, E., 2005, *A Preliminary Basin Profile of the Orange-Senqu River.* Pretoria: University of Pretoria.

Heyns, P. S., Patrick, M. J., and Turton, A. R., 2008, 'Transboundary water resource management in Southern Africa: meeting the challenge of joint planning and management in the Orange River basin', *International Journal of Water Resources Development,* 24(3): 371–383.

IMD, 10 June 2013, 'Why the Lesotho Highlands Water Project is a great idea', accessed August 2013 at http://www.interactmedia.co.za/index.php?option=com_content&view=article&id=4621:why-the-lesotho-highlands-water-project-is-a-great-idea&catid=230:plumbing-regulars&Itemid=323

IOL, 15 September 2011, 'Migration to shores heighten disaster risks', accessed August 2013 at http://www.iol.co.za/dailynews/opinion/migration-to-shores-heighten-disaster-risks-1.1137756#.UgoQ3JIm-So

Kistin, E. J., and Ashton, P. J., 2008, 'Adapting to change in transboundary rivers: An analysis of treaty flexibility on the Orange-Senqu River basin', *International Journal of Water Resources Development,* 24(3): 385–400.

Knoesen, D., Schulze, R., Pringle, C., Summerton, M., Dickens, C., and Kunz, R., 2009, *Water for the Future: Impacts of Climate Change on Water Resources in the Orange-Senqu River Basin.* Pietermaritzburg: Institute of Natural Resources.

Kranz, N., and Vidaurre, R., 2008, *Institution-Based Water Regime Analysis Orange-Senqu: Emerging River Basin Organisation for Adaptive Management.* Berlin: NeWater.

Kranz, N., Menniken, T., and Hinkel, J., 2010, 'Climate change adaptation strategies in the Mekong and Orange-Senqu basins: what determines the state-of-play?' *Environmental Science and Policy,* 13: 648–659.

LHWC, 2009, *Main Report: Feasibility Study for Phase II of the LHWP.* Maseru: LHWC.

LHWP, 2011, 'Water and electricity sales historical data', retrieved 2012 from Lesotho Highlands Water Project: http://www.lhwp.org.ls/Reports/PDF/Water%20Sales.pdf

Mail & Guardian, 25 February 2011, 'Water, not energy, shortage will constrain growth', accessed 9 August 2013 at http://mg.co.za/article/2011–02–25-water-not-energy-shortage-will-constrain-growth

NBI, 2011, *Component 3 Report: River Basins Organizations Survey.* Entebbe: Nile Basin Initiative.

ORASECOM, 2000, *Agreement between the Governments of the Republic of Botswana, the Kingdom of Lesotho, the Republic of Namibia and the Republic of South Africa on the Establishment of the Orange-Senqu River Comission.* Centurion: ORASECOM.

ORASECOM, 2011, *Projection of impacts and Guidelines on Climate Change Adaptation.* Centurion: ORASECOM.

Public Eye, 26 April 2013, 'Changing lives through climate change', accessed August 2013 at http://publiceye.co.ls/?p=2060

Quibell, G., 16 August 2013, former project officer for ORASECOM EU project, A. Earle, Interviewer.

Robbroeck, T., 2007, 'Water on the brain', *Civil Engineering,* 16(6): 11–19.

Schuermans, A. M., Helbing, J., and Fedosseev, R., 2004, *Evaluation of Success and Failure in International Water Management: the case of the Orange-Senqu River.* Zurich: ETH.

The Namibian, 29 April 2013, 'Climate change: Expert says Africans should be pro-active', accessed at August 2013 at http://www.namibian.com.na/indexx.php?archive_id=107626&page_type=archive_story_detail&page=1

Turton, A. R., Meissner, R., Mampane, P. M., and Seremo, O., 2004, *A Hydropolitical History of South Africa's International River Basins.* Pretoria: Water Research Comission.

UNDP-GEF, 2008, *Preliminary Transboundary Diagnostic Analysis of the Orange-Senqu River Basin.* Pretoria: UNDP.

VNJIS, 1992, *Agreement on the Vioolsdrift and Noordoewer Joint Irrigation Scheme.*

WWF-SA, 8 August 2013, *WWF South Africa climate change overview,* accessed 2013 at http://www.wwf.org.za/what_we_do/climate_change/

8 Conclusion

Climate change spanning from weather to water governance

As was stated in the introductory chapter, the aim of this volume is to increase the understanding of climate change impacts as it has been perceived, understood, and acted upon in a number of transboundary river basins globally. A particular (but not exclusive) focus was on the responses of the transboundary water management (TWM) institutions in the face of the climate change debate. Below, we will first revisit the Global Climate Change (GCC) debate, before we turn to the findings of our cases, from which we reinsert our findings into the debate on the nexus of TWM and GCC. The chapter, and the volume, is concluded offering a brief reflection on urgent research to be pursued in this field.

Evolution of the climate change debate and the relation to TWM

At the outset of this volume we remarked on the way weather-related climatic events have entered the public discourse, through extensive media coverage of disaster events as well as individuals ascribing extremes of weather to climate change. This process has continued, with climate now moving beyond being a phenomenon expressed through weather patterns through to a, contested, political reality. Indeed in February 2014, southern parts of Britain were threatened by unusually high floodwaters, the result of heavy rains and built-up floodplains. Politicians rushed to the scene of the flooding and sought to underline how, if their predictions had been listened to, the current situation could have been averted. *The Economist* magazine carried the headline 'Knee-deep in floodwater, Britain's politicians rekindle an argument about global warming' (Bagehot's notebook 2014), in an article chronicling the way the climate change debate has found its way into mainstream politics. As the saying goes, 'all politics is local', hence it is not surprising to find that the entry point for the global climate change debate into the popular political sphere is the very direct impact a weather-related event has on local communities. These impacts require that politicians act, to protect communities as well as to safeguard their own positions and popularity.

Scientists have been predicting human-induced climate change and its potential consequences for the last 200 years. In 1827, French scientist Jean-Baptiste

Fourier identified the warming effects of increasing greenhouse gases. Towards the end of 19th century, Swedish chemist Svante Arrhenius was the first one to calculate the effect of increasing accumulations of carbon dioxide in the earth's atmosphere (Houghton 1994: 12), though he was arguing that the increasing global warming would lead to better climate for northern European countries. In the last century, several scientific assessments were made about the potential disastrous consequences of human–induced climate change. However, only in 1998, the hottest year in over 100 years, the climate change issue was introduced to public discourse for the first time (Christianson 1999). As Leiserowitz argues, 'Public opinion is critical because it is a key component of the socio–political context within which policy makers operate. Public opinion can fundamentally compel or constrain political, economic and social action to address particular risks' (Leiserowitz n.d.: 3). Though climate change has become a mainstream political issue in many parts of the world, the public discussion on climate change, its causes and impacts on transboundary water sharing, continue to be prejudiced and incongruent. The clash between the climate change skeptics and the mainstream scientific view continues, with each continuing to undermine the other's positions. Skeptics constitute a very small minority, but they cannot be ignored, since their view has decisive influence on the public who is generally considered unwilling to change its lifestyle or moderate its aspiration.

In order to effectively adapt to global climate change challenges on shared water issues, global cooperation is a must. However, climate change has been among the most controversial topics in the recent history of global multilateral negotiations. GCC discourse has met a lot of resistance from certain actors, namely because of its links with the strong sectors of industry and energy. Critical voices are plentiful, coming from the private and business sectors, but also from the scientific community, giving voice to vested interest and/or inbuilt conservative views (as in the scientific community). We may expect anything but a smooth adaptation.

Predicted impacts of GCC on water availability pose serious survival threats to all people and institutions, but the policy response to that threat differs significantly across countries. Some countries are willing to engage with politically risky and economically expensive climate change adaptation policies, while others pretend to deny even the occurrence of climate change or just keep on blaming others while doing nothing at home. There is a wide variation in adapting effective climate change policies even among the industrialised developed countries. Thus, it is a very fundamental question to ask, theoretically as well as for its policy value: Why do some countries try to formulate and implement effective climate change adaptation policies while others do not? Is it only because of economic reasons, or does the role of political leadership and power of public opinion matter? There are also several countries that have already introduced climate change adaptation policies. But to measure and compare their effectiveness is not easy. As such, there is no golden rule on what constitutes the appropriate adaptation policy for a specific climate change challenge across countries and regions. Moreover, adaptation to climate change is very much a context-specific and local process, thus locating successful adaptation processes demands context-specific knowledge. However, it is not the formulation of the policy but how effectively it gets implemented that is the key. Climate change adaptation policy may not be properly implemented due to institutional ineffectiveness. This does not always mean that the state is shirking from climate change effort or subverting it (Steves & Teytelboym 2013).

Climate change is best projected as a threat multiplier, which aggravates existing social tensions and political instability (Swain et al. 2011). The real challenge is that climate change is likely to add additional pressures on the states and regions that are already fragile and conflict prone. In this context, it is crucial to investigate what influence the risks and challenges of climate change have on the transboundary river basins in Asia, Africa, and the Middle East, and how they should be dealt with policy wise. Climate changes are predicted to pose serious risks of decreasing agricultural production, widespread water shortage, declining food and fish stocks, increased flooding, and severe droughts. The transboundary river basins in Asia, Africa, and the Middle East, which we have discussed in detail in this book, are extremely vulnerable to possible climate change effects, a situation further aggravated by the interaction of multiple stresses these regions already face and their overall low adaptive capacity. Successful adaptation to this emerging threat is of crucial importance for political and socio-economic stability and future economic development in these regions. However, adaptation to the effects of climate change is a major challenge for highly vulnerable poor riparian countries in these basins. Their adaptation strategies cannot be successfully developed in isolation. All the necessary stakeholders in the basin are needed to engage in partnership. However, success of an adaptation strategy should be measured in a much more comprehensive manner. As Adger, Arnell, and Tompkins (2005: 77) argue, 'elements of effectiveness, efficiency, equity and legitimacy are important in judging success in terms of the sustainability of development pathways into an uncertain future'.

While taking a casual approach on the adaptation front, several riparian countries, such as China, Israel, and Egypt, are encouraged to turn the climate change debate into a debate about security. Securitisation of possible climate change impact on trans boundary river water sharing is not necessarily bad, as it sometimes lifts the issue into the agenda of high politics where it may get the required political attention. However, the possible implication of this securitisation process is precisely that it prevents broad stakeholder participation in prioritisation and decision-making. Securitisation might also prompt military solutions to transboundary water issues, and the basin states' focus on military responses might adversely affect the preferable alternative strategy of adaptation (Brown et al. 2007). When the military takes the center stage in climate policy-making, attention would be distracted away from the need for wider socio-economic changes and technological developments.

Some basin countries are even using the 'climate change debate' to justify their unilateral water development policies in transboundary river basins. For instance, the new priority of the Chinese government to build large hydro-power dams in the ecologically and ethnically sensitive Nu–Salween River basin in the name of reducing its dependence from fossilised carbon based energy production and thus cutting down its carbon dioxide emissions. These actions lead to subversion of the process of arriving at a basin based coordinated approach in the face of GCC challenges. China not only takes the help of the climate change debate to justify its large dam buildings on Nu-Salween, it also adopts the same route of 'subversion' while justifying the unilateral operation of its large dams on the Lancang-Mekong River. In one of the Mekong River Commission (MRC) meetings in Thailand, 2010, China was criticised by Laos, Thailand, Vietnam, and Cambodia, who blamed Chinese upstream dams for causing downstream floods in rainy seasons and droughts in the dry-season. Chinese delegates, how-ever, blamed the climate change-induced variation in precipitation as the cause for floods and droughts in the downstream. In the Nile basin, while Ethiopia is justifying its ongoing construction of the Grand Renaissance Dam as it will help meet climate change-induced river flow variation challenges, Egypt takes the same climate change debate to oppose the project, saying that its downstream impact in the future cannot be properly assessed. Thus, it is important to criti-cally analyse the use and misuse of the climate change debate by the riparian countries in a transboundary basin setup. This issue is still pressing and it is necessary to analyse it in a systematic manner – a first attempt at which this book represents.

Learning from our six cases – adapt, resist, subvert

In this volume, six rather disparate cases have been examined regarding their responses to the emerging debate on GCC. The six cases represent a range of geographical types and institutional setups. Their distinct political economies have meant that the challenges experienced in each have been addressed in somewhat different ways. The six different cases have all been subjected to a framework developed for analysing respective basin (and riparian) under the Adapt (A), Resist (R), Subvert (S) headings. In Chapter 1, the analytical concepts were presented and described as: *Adapt*: by adopting climate change-oriented concepts, promoting legal instruments, establishing institutional frameworks, and implementing action plans at the regional/basin, national, and even local level; *Resist*: by refusing or contesting the fundamentals of the climate change discus-sion, and deliberately disregarding it in the policies and discourses adopted in the management of transboundary water resources; *Subvert*: by using the climate change discussions' concepts, legal, and institutional frameworks for purposes other than to adapt/mitigate the climate change impacts.

River basins are different, and hardly lend themselves to a strict comparative analysis, and even less on quantitative grounds, as there are globally too few cases with too many variables. Having said that, the framework below can still be used to structure the data from the various cases. The responses are typologised in the *adapt-resist-subvert* triology that has been used to analyse each case. In this concluding chapter, we contrast the responses with one another in order to tease out similarities and differences between basins and between the various types of responses (see Table 8.1). The framework allows for an assessment of the various responses within each case (horizontal comparison), but also to assess how the various responses appear in different basins (vertical comparison).

While the cases are different, there are also similarities between them. Some basins (primarily the Jordan, the Ganges-Brahmaputra, and to some extent the Nile basin) are clearly more securitised (to speak with Busan et al. 1998), while the others are more in a 'development mode' (the Mekong, the Oranqe-Senqu, the Niger, and to some extent the Nile basin). These differences are also evident in the way the various discourses at a general level as well as in relation to climate change is developing within each basin. As noted above, the institutional frameworks in the different basins are very different. Ranging from largely non-cooperation or very low degree of cooperation (the Jordan basin or the Ganges-Brahamaputra basin) to basins where cooperation is slowly being institutionalised (the Nile basin, the Niger basin) to basins where cooperative institutions/structures have been in place (in one shape or form) for a longer period (the Mekong and the Orange-Senqu). This is to a rather large degree an outcome of the prevailing and historical political economies of the basins/regions. For the Jordan basin and the Ganges-Brahmaputra basin, the overarching level of conflict and focus on security considerations have been the major factor.

Turning to the analytical framework (see below), reading it horizontally, we can see that in the *Ganges-Brahmaputra* basin with its high stakes in water, high population pressure, a strong hegemon (India), and the unresolved historical conflicts, the riparians have failed (hegemon has chosen to fail) to establish basin-authorities that could have responded forcefully to GCC. The *Jordan* basin shares these characteristics of having no basin authority and a conflict-ridden context, drowning out regional responses. Instead, both basins look for water elsewhere and whatever preparation for GCC that is going on is on national level. New water (diversion and desalinisation, respectively) is the favoured solution, sometimes justified by GCC. While not overtly *resisting*, by neglecting to act, the net effect remains the same.

In contrast, *Mekong* and *Niger* have river basin authorities where the GCC plays a considerable role and operates in a milieu where water is only moderately securitised. Whereas the Mekong basin experiences growth and development and, therefore, has some development space, the Niger has deep-seated development problems which at large take over from the GCC-dynamics, in part creating a *subversion* of the responses to the GCC. MRC in Mekong has the ability to respond to GCC challenges, but the comprehensive and well-funded program dealing with this is stalled and is far from delivering what it was supposed to.

The last 'pair' is not a symmetric pair at all: The *Nile* is a huge system, utterly complex to govern. There is a range of infrastructure projects claimed to *adapt* the Nile to the GCC requirements, but most are only thinly related to GCC. There is, moreover, a fair amount of *subversion* where GCC is used to further other interests. Overall, transboundary cooperation has so far failed

Table 8.1 Analytical framework for the ARS-approach

	Adapting	*Resisting*	*Subverting*	*Comment*
Ganges–Brahma	Very little adaptation.Minor adaption on national level.	Lack of meaningful basin cooperative structures at regional/basin level	India: aiming to divert water from other sources instead of accepting the GCC constraints.	*Responses suffering from regional conflict and lack of multilateral basin cooperation*
Jordan	Adaptation on national level, overlapping with prior issues (e. g. desalination)	Indirectly: lack of cooperative structures	Intermingles with and is 'hidden' within deep-seated regional conflicts. Desalinisation is the 'new water'.	*Responses suffering from high regional conflict level and complete lack of multilateral basin cooperation*
Mekong	Ongoing basin level programs for preparing for GCC. National level preparations through NAPAs is ongoing. Donor-driven.	No overt resistance. Sluggishness. Strategic ignorance.	Lip-service adaptation at basin level. National level activities, but uncoordinated with basin level.	*The RBA and the donors engages, but is already struggling with inability to pursues its programs*
Niger	GCC an active part of RBA-development. Donor-driven.	Indirectly: drowning in deeper problems of water and energy supply	Pursuing GCC responses under previously defined dam and irrigation plans.	*GCC responses in Niger are overlapping almost perfectly with previous development dilemmas.*
Nile	Infrastructure projects pursued in name of responses. Adaptation on national level goes on. Donor-driven.	Too obvious to resist. Indirectly through the incomplete basin authority.	GCC is utilised for furthering vested financial and political issues.	*GCC as smoke-screen for 'failed' transboundary cooperation.*
Orange	Shallow adaptations attempts.	Ignoring GCC, letting other priorities rule.	No major attempts at subverting the GCC into other agendas.	*Powerful interests and reasonable consensus on current regime serve to neglect responses to GCC*
Comment	*Adaptation tend to be pursued:* • *At national level* • *Mostly rhetorically* • *As a part of existing interventions* • *Only if there is a functioning RBA*	*Resistance tend to be indirectly present through:* • *Dysfunctional RBAs* • *Deeper prior problems/conflicts* • *Strategic ignorance*	*Subversion is pursued in many ways, such as:* • *Becoming an argument for "standard" modernisation projects* • *Redirected to be an argument in a political conflict* • *Manipulation of national vs regional interests.*	

in the Nile, which also explains why GCC responses are not so strong. *Orange* has a fairly well-established river basin regime and is largely governed by middle-income countries. Here, only shallow attempts at *adaptation* are pursued, but at the same time there is neither overt *resistance*, nor *subversion*. Instead concrete business plans on water and energy are untouchable, rendering GCC marginal for now.

Turning to a *vertical reading* of the framework, we conclude that *adaptation* to GCC takes place predominantly in those cases where there is a basin authority in place, and often enmeshed with already ongoing interventions and/or projects that also could be legitimised from other development/security points of view. In this, our findings coincide at large with conclusions from a previous study on GCC-adaptation in Africa, stating 'adaptation in the water sector is beginning to emerge, although evidence suggest this is primarily in the form of building adaptive capacity and no-regret type activities in response to multiple factors' (Pahl-Wostl 2009: 823). As a way of critique, the lion's share of adaptation takes place on national level, with little regard for regional concerns, calling for more attention to a 'conceptual framework for analysing adaptive capacity and multi-level learning processes' (354). Sometimes, even worse, the talk on responses remains rhetorical. Many times adaptation is not so much adaptation as it is business as usual in new forms, or even no business at all, just words.

For the next category, *resistance*, we find that it is rarely occurring in a pure form. As such the attitudes in the basins do not reflect the global scientific debate on whether GCC is happening or not. More subtle forms of resistance could nevertheless be traced. Firstly, the absence of a proper basin authority, tend to fragment understanding and governance, hence paralysing efforts at responding to GCC. In this form, resistance is rarely articulated, but the result could be interpreted as resistance to a systematic response. Secondly, in some basins, there are violent and protracted conflicts, in others acute development needs that by necessity push GCC down the ladder of priorities. Thirdly, in some places we have detected what could be named 'strategic ignorance', implying that GCC is, of course, known, but decision-makers prefer to stick the head in the sand because any other choice would be politically and financially uncomfortable (or even infeasible).

Finally, *subversion* may be the most interesting, and definitely the most frequently seen outcome. As expected, GCC introduces change in a politicised/securitised field, which various interests are keen to capitalise on. Also predictably, standard modernisation projects (like dambuilding or irrigation) are typically now dressed in new colours and presented as imperatives for preparing for GCC, and to which fresh resources can be drawn. On the one hand this is a construction, but on the other it cannot be dismissed easily since they may very well make sense (in spite of being a post-facto construction). In a similar fashion, GCC arguments are commonly transformed into fitting arguments in political conflicts, sometimes far detached from this discourse. Finally, there is a subversion of sorts going on where national activities, interests, and priorities are 'played' vis-à-vis regional/basin activities, interests, and priorities. This is related to the collective action dilemma (Olson 1965) where all want 'the other' riparians to serve a general interest, while they serve their own interest. It is evident that the GCC argument is largely invoked when it is perceived as useful (in terms of providing a legitimate argument for the political interest), and dismissed when it is not.

So trying to pull out the core insights from our cases and our framework, we would end up with four related themes. *Basinwide institutions/governance bodies/ RBAs* provide an arena and a mechanism within which to address issues of regional importance such as climate change. They are necessary for a concerted effort at addressing GCC, but not sufficient. Cases with RBAs show a mixed result, but those without it have very weak GCC responses. This is true for the discourse on climate change in each of the Niger, Nile, Orange-Senqu, and, especially in the Mekong basin, on the one hand, and on that in the Jordan and Ganges-Brahmaputra on the other.

Furthermore, the research shows that most of the actual adaptation happens at the *national rather than the regional level.* To mention the most obvious example, the Nile basin climate change is in focus in many of its projects and programmes. Sometimes GCC responses also take on political dimensions as one important way to address climate change challenges. Infrastructural works can enable countries to better meet the challenges of climate variability and change. In this, national concerns are typically pursued, rather than regional ones, which in the worst-case scenario will disrupt rather than contribute to basin cooperation and to a comprehensive work with preparing for GCC. It is noteworthy, however, that donor preferences for funding activities on the regional level seem to have an important impact on the allocation of adaptation measures in the most-impoverished basins.

It is clear that the prevailing discourse in the respective basins in many regards sets the boundaries for what is *politically feasible* in relation to action on GCC. Again, to illuminate with the most obvious example, in the Jordan basin a truly regional framework to address the issue is not politically possible because of the Arab-Israeli conflict, hence undermining possibilities of a response harmonised at basin level. In the Mekong, the so-called Mekong spirit may have served to give a great space for initiating cooperation on climate challenges. GCC and water issues are clearly subordinate to higher political priorities, where security is the main obstacle and previous cooperation is useful.

In yet other cases, the focus on GCC has largely come out of an increasing focus by *donors/development partners* on the topic. Needless to say, it has provided opportunities for financing of projects on climate change adaptation in relation to TWM transboundary water management, but also at the national levels. Civil society has also played a role in highlighting GCC in the respective basins, but it is generally not gaining so much influence when the formal structures are not susceptible to it. This has in our cases, however, only to a limited degree affected the outcome and depth of GCC responses.

Concluding thoughts on findings in context

TWM is in flux, and there is a wide consensus that GCC will force a change: 'The impacts of climate change and other stresses on water resources and changes to flooding risks in the future will require adaptation on the part of water resource management institutions and water users' (Goulden et al. 2009). The presence of functioning international institutions may be of utter importance for the development of resilient social ecological systems that manage to deal with changes in climatic variability (De Stefano et al. 2012). Finally, international institutions are also considered essential in intervening in the relationship between climate change and security (Tir & Stinnett 2012: 212). Critically, these severe statements are often accompanied by calls for innovative and adaptive ways of governing water, for which current institutional arrangements are considered faltering (Huntjens et al. 2012: 67). Pahl-Wostl sums it up eloquently by stating that adaptive management – which we need to adopt in an era of uncertainty – is essentially about 'learning to manage by managing to learn' (Pahl-Wostl 2007: 58). The contemporary empirical evolution around these uncertainties is what this volume aims to illuminate. As in many other sub-fields of TWM, this is to be found in the design and quality of the institutions set to govern, and the degree and nature of cooperation among the riparians.

From a theoretical perspective, the management of transboundary waters remains a collective action challenge. Mancur Olson (1965) outlines the main facets of collective action theory, inadvertently explaining non-cooperation. This is so since actors have a tendency to try to 'free-ride' when there is an opportunity to do so. While still 'cooperating', they would tend to avoid sharing the costs (as in financial and other adaptive burdens), often de facto disrupting substantial cooperation. Notwithstanding that violent conflicts may be kept at bay, 'commons' are exploited, over-used, gradually producing a 'tragedy' of sorts (Hardin 1968). Hence, the reason for, and quality of, cooperation is often more important than the simple lip-service frequently paid to shallow 'cooperation' (Yoffe et al. 2003) resulting from collective action failure in rigid but non-committing agreements (Mirumachi 2010). Some of the more efficient transboundary water governance processes are instead based on shared knowledge, informal institutions, and regime formation (Young 1982; Krasner, 1983; North 1990; cf. Ostrom 1990). While solutions are not easily found, the collective action problem is reasonably well defined.

Moreover, taking into account that we in this volume analyse the climate change debate (and not the climate change itself), and how this debate has taken hold and been used/applied in various basins, yet another dimension of change appear. Knowing that TWM-arrangements often are influenced by globally sanctioned discourses,[1] how do they fare when another sanctioned discourse (i.e. GCC) arrives? Hypothetically, actors may deliberately emphasise (and/or de-emphasise) certain perspectives according to vested political interests (Jägerskog 2003; Warner & Zeitoun 2008: 127). Powerful actors are usually those with strongest capacities to sanction discourses and best able to disseminate widely its preferred perspectives while discarding alternatives. Thus, GCC opens up for power being projected discursively on a higher level than before, which is likely to be used by those benefitting from it.

Responding to GCC adds to the mix of TWM and to some extent changes the rules of the game. *First*, as was pointed out in Chapter 1 and in many case studies, the agreements which are founded in numerical water allocation, mean flows, annual precipitation, and/or run off measured for decades (or centuries) in a comparatively transparent manner are now increasingly questioned. The dimension of change (now or soon) enters the scene, triggering the re-opening of long-since concluded negotiations for which many agreements are ill-prepared (cf. Fischhendler 2004). *Second*, even if water agreements do not mention figures, and even if there is no proper agreement, the common understanding and the shared knowledge (or the agreed disagreements) of the water regime is nevertheless the base for the regime at hand. GCC serves potentially to alter both tangible water availability/access and discursively serve to disrupt common understandings in the various basins. Hence, it deserves to be taken seriously.

The governance dilemma of uncertainty and variability in water access in transboundary basins is enjoying growing attention. In particular, the school of thought loosely collected in the 'adaptive management' approach has been convincing, and has to some extent been developed in relation to GCC (Pahl-Wostl 2007, 2009; Pahl-Wostl et al. 2008; Cots et al. 2009; Goulden et al. 2009; Huntjens et al. 2012). The key argument is to develop the societal and institutional capacity to learn and to adapt, and especially to focus on information sharing/gathering on the one hand, and on the ability to process that information collectively and constructively, on the other (Pahl-Wostl 2007: 52). However, as Pahl-Wostl also points out, and as is abundantly clear from most of our cases in this volume, politicised transboundary basins in water scarce areas do neither lend themselves easily to any smooth information sharing and learning, nor are their key actors keen to constructively adapt their policies for the common good. The insights here resonate with the claims of Goulden et al. (2009) that most of the research on GCC adaptation has been made on river basins in the United States and in Europe, as well as the argument of Öjendal, Hansson, and Hellberg (2012) that much of the progress in water management the last decade is designed for less-politicised and better-resourced contexts. Or as Goulden, Conway, and Persechino (2009: 805) say, 'without appropriate cooperation adaptation may be limited and uneven'; appropriate cooperation is rare in deeply contested waterscapes, hence adaptation suffers.

However, as we have seen, responses to GCC in transboundary basins are not quite the burden that many other adaptations, compromises, and growing scarcities amount to. In several of our cases, GCC has added to the construction of institutional governance bodies, and in all cases, GCC serves to attract resources to the basin and the riparian countries; i.e. the international community are often sympathetic to contributing resources for adaptation-measures and/or for other (already existing) projects that are revised/re-defined as being relevant for responding to GCC. As such, the GCC constitutes an asset, a tool which eventually will serve to attract resources to solve old or new problems. We also see that there is a connection – although elusive and indirect at times – between basins with a governance body and the ability to attract resources for the purpose of responding to GCC. In this light, GCC can also be seen as an avenue to improve regional cooperation and governance efficiency.

Recommendations for further research

Basin cooperation seems to be a key factor for the ability to adapt to the demands of the emerging GCC and its localised consequences. Cooperation has been the focus for a long time in an effort to enhance development efficiency and to avoid disruptive conflicts (Yoffe et al. 2003; Mirumachi 2010). However, which dimensions of cooperation are central for the particular cause of enhancing GCC-adaptation? Are there particular emphases and designs necessary for institutions to be able to optimally respond to GCC challenges?

Theoretically (Pahl-Wostl 2009), as well as empirically (as seen in this volume), *multilevel learning processes*, including institutional design and nature of cooperation, are imperative for creating capacity to adapt to uncertainties. How would such learning processes look like? What would be needed to initiate them?

Finally, as we have seen, there are mild (and sometimes not so mild) *subversive activities surrounding the accelerating support for adaptation to GCC* in the various basins. These are often tied to prior projects and existing programmes, sometimes with little alteration from its original idea. While this may be legitimate, there is also a certain 'gliding' along the deep and wide uncertainties that come with the GCC territory. This is likely to both eschew focus and to derail resources into standard modernisation projects. How will this be identified and assessed?

In closing, the GCC has opened up an important debate on how to better the imperfect practices of transboundary management of common water resources under a regime of both scarcity and uncertainty. This promises to be a complex yet urgent undertaking for the long-term for researchers and practitioners alike.

Note

1 Referring to the prevailing or dominant ideas and narratives that are discursively and politically legitimised by certain actors, groups, or states is the notion of sanctioned discourse (Gramsci 1971; Allan 2003).

References

Adger, Neil. W., Arnell, Nigel W., and Tompkins, Emma L., 2005, 'Successful adaptation to climate change across scales', *Global Environmental Change,* 15(2): 77–86.

Allan, J. A., 2003, 'IWRM/IWRAM: A new sanctioned discourse?' Discussion Paper No. 50. Water Issues Study Group. London: University of London.

Bagehot's notebook, 22 February 2014, 'Whither the weather? Knee-deep in floodwater, Britain's politicians rekindle an argument about global warming', *The Economist,* accessed

16 February 2015 at http://www.economist.com/news/britain/21596933-knee-deep-floodwater-britains-politicians-rekindle-argument-about-global-warming-whither

Brown, Oli, Hammill, Anne, and Mcleman, Robert, 2007, 'Climate change as the "new" security threat: implications for Africa', *International Affairs* 83(6): 1141–1154.

Buzan, Barry, Wæver, Ole, and Wilde, Jaap de, 1998, *Security: A New Framework for Analysis.* Boulder, CO: Lynne Rienner.

Christianson, G. E., 1999, *Greenhouse: The 200-Year Story of Global Warming.* New York: Walker and Company.

Cots, F., Tàbara, J. D., McEvoy, D., Werners, S., and Roca, E., 2009, 'Cross-border organisations as an adaptive water management response to climate change: The case of the Guadiana river basin', *Environment and Planning C: Government and Policy,* 27(5): 876–893.

De Stefano, Lucia, Duncan, James, Dinar, Shlomi, Stahl, Kerstin, Strzepek, Kenneth M., and Wolf, Aaron T., 2012, 'Climate change and the institutional resilience of international river basins', in *Journal of Peace Research,* 49: 193.

Fischhendler, Itay, 2004, 'Legal and institutional adaptation to climate uncertainty: a study of international rivers', in *Water Policy,* 6(4): 281–302.

Goulden, M., Conway, D., and Persechino, A., 2009, 'Adaptation to climate change in international river basins in Africa: a review', *Journal des Sciences Hydrologiques,* 54(5): 805–828.

Gramsci, A., 1971, *Selections from the Prison Notebooks.* New York: International Publishers.

Hardin, Garrett, 1968, 'The tragedy of the commons', *Science* 162(3859): 1243–1248.

Houghton, J., 1994, *Global Warming: The Complete Briefing* (2nd ed.). Cambridge, UK: Cambridge University Press.

Huntjens et al., 2012, 'Institutional design propositions for the governance of adaptation to climate change in the water sector', *Global Environmental Change,* 22(1): 67–81.

Jägerskog, A., 2003, *Why states cooperate over shared water: The water negotiations in the Jordan River basin,* PhD Dissertation, Linköping University, Sweden.

Krasner S., 1983, 'Structural causes and regime consequences: regimes as intervening variables', in Krasner, S. (ed.), *International Regimes,* Ithaca, NY: Cornell University Press.

Leiserowitz, Anthony, n.d., 'International public opinion, perception, and understanding of global climate change', accessed 16 February 2015 at http://environment.yale.edu/climate-communication/files/IntlPublicOpinion.pdf

Mirumachi, Naho, 2010, *Study of conflict and cooperation in international river basins,* PhD thesis, London School of Economics.

North, Douglass C., 1990, *Institutions, Institutional Change and Economic Performance.* Cambridge, UK: Cambridge University Press.

Öjendal, Joakim, Hansson, Stina, and Hellberg, Sofie, (eds.), 2012, *Politics and Development in a Transboundary Watershed – The Case of the Lower Mekong Region.* London: Springer.

Olson, Mancur, 1965, *The Logic of Collective Action. Public Goods and the Theory of Groups.* Cambridge, MA: Harvard University Press.

Ostrom, Elinor, 1990, *Governing the Commons – The Evolution of Institutions for Collective Action.* Cambridge, MA: Harvard University Press.

Pahl-Wostl, 2007, 'Transitions towards adaptive management of water facing climate and global change', *Water Resources Management,* 21: 49–62. Doi: 10.1007/s11269-006-9040-4.

Pahl-Wostl, 2009, 'A conceptual framework for analysing adaptive capacity and multi-level learning processes in resource governance regimes', *Global Environmental Change,* 19: 354–365.

Pahl-Wostl, C., Gupta, J., and Petry, D., 2008, 'Governance and the global water system: a theoretical exploration', *Global Governance* 14(4): 419–435.

Steves, Franklin, and Teytelboym, Alexander, 2013, 'Political economy of climate change policy', Smith School of Enterprise and the Environment School of Geography and the Environment, University Of Oxford, Working Paper 13-02, October 2013.

Swain, Ashok, Bali Swain, Ranjula, Themnér, Anders, and Krampe, Florian, 2011, *Climate Change and the Risk of Violent Conflicts in Southern Africa*, Pretoria: Global Crisis Solutions.

Tir, Jaroslav and Stinnett, Douglas M., 2012, 'Weathering climate change: Can institutions mitigate international water conflict?' *Journal of Peace Research*, 49: 211. doi: 10.1177/0022343311427066.

Warner J. F., and Zeitoun, M., 2008, 'International relations theory and water do mix: a response to Furlong's troubled waters, hydro-hegemony and international relations', *Political Geography*, 27: 802–810.

Yoffe, S. B., Wolf, A. T., and Giordano, M., 2003, 'Conflict and cooperation over international freshwater resources: indicators of basins at risk', *Journal of the American Water Resources Association*, 39(5): 1109–1126.

Young, Oran R., 1982, 'Regime dynamics: the rise and fall of international regimes', *International Organization International Organization*, 36(2): 277–297.

Index

Page numbers in *italics* indicate figures and tables.